# Just the Facts
## Investigative Report Writing

# Just the Facts
## Investigative Report Writing

**MICHAEL BIGGS**
Long Beach City College

**Prentice Hall**
**Upper Saddle River, New Jersey 07458**

Library of Congress Cataloging-in-Publication Data

Biggs, Michael, 1950–
   Just the facts : investigative report writing / Michael Biggs.
        p.   cm.
   Includes index.
   ISBN 0-13-014301-4
   1.  Police reports--Handbooks, manuals, etc.   2.  Report writing--Handbooks, manuals,
etc.   3.  Criminal investigation--Handbooks, manuals, etc.   I. Title.

   HV7936.R53 B54 2000
   808'.066363--dc21                                                                99-088616

Senior acquisitions editor: *Kim Davies*
Production editor: *Patricia Noble*
Production liaison: *Barbara Marttine Cappuccio*
Director of manufacturing and production: *Bruce Johnson*
Managing editor: *Mary Carnis*
Manufacturing buyer: *Ed O'Dougherty*
Creative director: *Marianne Frasco*
Cover design: *Bruce Kenselaar*
Cover Illustration: *Timothy Cook/Stock Illustration Source*
Interior Illustrations: *Michael Barr*
Marketing manager: *Shannon Simonsen*
Editorial assistant: *Lisa Schwartz*
Formatting/page make-up: *Clarinda Company*
Printer/binder: *Banta Book Group*

© 2001 by Prentice-Hall, Inc.
Upper Saddle River, New Jersey 07458

Printed in the United States of America

10  9  8  7  6  5  4  3  2

ISBN 0-13-014301-4

Prentice-Hall International (UK) Limited, *London*
Prentice-Hall of Australia Pty. Limited, *Sydney*
Prentice-Hall Canada Inc., *Toronto*
Prentice-Hall Hispanoamericana, S. A., *Mexico*
Prentice-Hall of India Private Limited, *New Delhi*
Prentice-Hall of Japan, Inc., *Tokyo*
Prentice-Hall Singapore Pte. Ltd.
Editora Prentice-Hall do Brasil, Ltda., *Rio de Janeiro*

To the KCBs and Tucker

# Contents

# Preface

Since 1980 I have been an instructor in several areas of law enforcement training and security management, with an emphasis in the field of investigative report writing. This has included course preparation and presentation at many levels including basic and reserve academy courses, advanced officer and investigator courses, supervisor updates, and community college classes. I have had the opportunity to talk with thousands of law enforcement and security officers, field trainers, supervisors, and managers. Through this experience I have learned that a common concern is the need for an effective entry-level report writing training guide that would explain some of the basics of an investigation and how to write about them.

*Just the Facts* was written to help fill a need for a training tool that combines some of the basics of investigation with the basics of report writing for entry-level students and academy recruits. Before investigators can write about what they have done, they must know something about how to perform their craft. The existing literature covers the ends of this spectrum quite well. Libraries and bookstores are well-supplied with books and manuals on how to investigate something, as well as on how to write. Very few if any, however, try to combine the two disciplines and bridge the distance between them. The need for this type of work is evidenced by the number of young police officers who have difficulty turning their preliminary investigative efforts into quality reports. This book was not designed to teach someone how to write. It was designed to help teach someone—who knows how to write—how to write a police report. It assumes the student brings a working knowledge of the English language to the learning experience and, as such, makes no attempt at being a grammar book.

This workbook is an attempt to meet the needs of report writing students by establishing fundamental guidelines for investigative reports through a set of rules that are easy to understand and apply in any situation. By following these rules each major component of investigative report writing can be broken down to its simplest form and examined for weaknesses. These weak points can then be corrected with immediate improvements made.

Since 1985 I have taught police report writing at the community college level. As part of my preparation and course development I have reviewed most, if not all, of the available texts and journal articles dealing with this subject matter. The majority of these writings address the need to simplify and professionalize the style and content of reports through the teaching of grammar and spelling, however, none establish a method or set of rules to do so. *Just the Facts* puts forth a set of guidelines or rules to help students work through any type of report writing problem. It also presents scenarios in which the student can apply the learned behaviors in report writing situations. In this part of the learning process the students are able to test their knowledge in exercises ranging from fill-in-the-blank questions to writing reports based on role playing situations.

All too often young investigators are described as being poor report writers because their reports are short, difficult to understand, or lack detail. I suggest that if investigators correctly and accurately write what they discover during their investigations, they are good report writers. If the report is lacking substance, the problem is not one of report writing but one of investigative skill.

Fourteen years of practice with these rules of writing and exercises—with continual feedback from students, police officers who have attended the class and put these techniques into practice, and other report writing instructors—have convinced me that this system works.

The chapters are designed to identify key learning points followed by an explanation and example of each. Each chapter has a short review, a set of exercises to build on the chapter learnings, and a ten-question quiz. Questions are a mixture of true/false, short answer, and multiple choice. They are designed to build confidence and reinforce the material just covered. Each chapter is devoted to a major component of the report writing process and builds on the previous learnings.

The text is based on the premise that in order to write police reports, the student needs to know something about investigations. As such the text begins with a discussion of investigations to give the student a basic foundation from which to build writing expertise. Other chapters include note taking, rules of narrative writing, describing persons and property, crime reports, arrest reports, issues in writing, search warrants, and dictating reports.

Over the years the material in this text has been modified and field tested many times in academy settings and at the community college level with the hope that one day it would be right. Whether or not it meets the expectations of all the members of the criminal justice system remains to be seen, but the intent to do so is there. No workbook like this comes from a single source and I want to thank those who gave their time reviewing the manuscript and providing feedback on the text. This includes Sergeant Richard Butcher, Huntington Beach Police Department; Justice Patricia Bamattre-Manoukian, California Court of Appeals, 6th District; Mark Colin, Chevron Oil Spill Coordinator; Judge Sarah Jones-Fuller, Municipal Court, West Orange County Judicial District, retired; Captain Ed McErlain, Huntington Beach Police Department; Investigator Clay Searle, Los Angeles Police Department, retired; Everett Teglas, Chevron Corporate Security, Latin America; and in memoriam to Sergeant Bob Moran, Huntington Beach Police Department. I also want to thank Police Chief Tim Grimmond, El Segundo Police Department, and Captain John Rees, La Habra Police Department for their assistance in providing the report forms used as examples in the text. Last, but not least, I want to acknowledge and thank the hundreds of police officers and students who took the time to give me their thoughts and comments on how to improve the text. I tried.

*Mike Biggs*

# 1 Investigation Basics

Who is an investigator?

What is an investigation?

The goal of an investigation

The steps in initiating an investigation

The who, what, where, when, why, and how of investigations

The qualities of a good investigator

Investigative reports do not just happen. They are the formal product of some event that has been brought to the attention of an investigative body or agency which, after a thorough examination, has been documented. Since a good portion of an investigator's time is often spent investigating these events before a report is written, it is logical that any report writing teaching instrument should also include some discussion about investigators and their investigations.

## WHO IS AN INVESTIGATOR?

Investigators are people who look into events or situations to find the facts about what happened. They ask questions and interview people, look at crime scenes, examine documents, collect evidence, develop informants, find stolen or missing property, and develop an understanding of what occurred after reviewing all the available information. Investigators are police officers, deputy sheriffs, security guards, firefighters, claims adjusters, private investigators, personnel specialists, and many other categories of people who are required to possess certain investigative skills and knowledge in order to perform their duties. There is no single set of life experiences or level of education that qualifies someone to be an investigator, but as we will see, there does seem to be some common ground for those who are successful.

## WHAT IS AN INVESTIGATION?

An investigation means different things to different people; however, for the purpose of investigative report writing, an investigation may be defined as a lawful search for things or people. In each case the goal is to find the truth. This definition applies to both criminal and administrative investigations.

The media and the war stories of hundreds of investigators have created a commonly held belief that law enforcement representatives are able to investigate anyone at any time for any reason. Few things could be further from the truth because a criminal investigation cannot begin until one of three things is present:

1.  A crime must have occurred.
2.  There must be a reasonable certainty that a crime has occurred.
3.  The investigator must be reasonably sure a crime is going to occur.

This means the investigation and report writing process could occur in one of two ways. The first possibility would begin with the occurrence of a crime, followed by an investigation, and ending with the solution being discovered and proved. In the other sequence, the investigation begins, then the crime occurs, and ultimately the solution is discovered and proved.

An example of the first instance would have a patrol officer who is driving down the street see a suspect run out of a bank with a gun in one hand and a bag of money in the other. Just as this takes place the officer receives a radio call describing a bank robbery at that location. Another example would be when an officer arrives at the scene of a call and sees a dead body with two gunshot wounds to the back of the head. A second person at the site shows the officer a gun and says that he saw someone shoot the person lying on the ground and then run away.

In the second instance an investigator may have an informant who has reported that a burglary is going to occur. The informant tells the officer the names of some of the suspects but does not know the location or exact time the burglary is going to take place. The officer would very likely begin the investigation with an attempt to identify the suspected parties and try to figure out where the crime was going to take place. These beginning steps are the start of an investigation. Another example of the second instance might occur if a department had several reports of stolen cars from a shopping mall—and set up a surveillance to catch those responsible. One day an officer on a stakeout sees a person walk up to a parked car, break the window and after modifying the ignition, drive away in the car.

The parameters of an administrative or noncriminal investigation are broader than those of a criminal matter. If the purpose of the investigation is not illegal, and the civil rights of those involved are not infringed on, an administrative or civil investigation may proceed. Examples of civil investigations might include a review of property tax records to find out who owns a particular piece of land, or to find out how many square feet a building contains. To complete any of these investigations, the information contained in public records is not only open to review but available for use.

The proprietary rights of an employer are often the basis for an administrative inquiry. This can include investigation into employees use and entitlement to benefits, the misappropriation of company products or supplies, and determining whether someone is creating a hostile environment in the workplace.

Regardless of which occurs first, the administrative investigation or the reason for it, the investigator must constantly be aware of all aspects of the environment. This means you have to monitor and adjust your methods to all the conditions, circumstances, and influences in the investigation. Are the actions you are taking going to impede the ongoing purpose of the company you are working for? Are the interviews you are planning to do going to disrupt the workforce or cause production to stop? If so, it may be appropriate to meet with the business managers responsible for the operations of your employer and resolve the possible negative outcomes before you proceed. Conducting the investigation appropriately and handling problems or issues at this point makes the report writing part of the job a lot easier.

Investigators must also recognize that persons who appear to be uninvolved may in reality be suspects or witnesses. They may have something to gain or lose from disclosing or withholding their involvement in the matter. The people you are talking to may not honor your request to keep the matter confidential. It is human nature to talk to friends and co-workers about the events in our lives. You should expect some of the people you talk with to tell others about your investigation and even about the conversations you have had.

The investigator must also be aware that evidence may be difficult to locate, incomplete, or latent at first glance. In order to make the most of every opportunity to preserve evidence, and insure a complete investigation, the investigator should always strive to involve evidence specialists at the earliest time. Advancements in technology are happening all the time, and some of them are able to get useful information from evidence that was previously worked and found to be of no value.

Depending on the circumstances and regardless of whether it is a criminal or administrative inquiry, there are some basic steps that all investigators can take to initiate an investigation. As each opportunity to investigate begins, the good investigator considers these guidelines.

1. Arrive at the place you are needed to begin the investigation. More than one serious crime has gone without an investigation beginning promptly because someone crucial to the case became sidetracked or unavailable while getting to the scene. Crashing on the way or stopping to write a minor traffic citation is generally considered bad form.

2. There may be people at the scene who are in need of help or assistance. Remember that in emergency responses, the first order of business is life safety, and a big part of any investigator's job is to help those in need. Helping someone in need can go a long way towards developing a spirit of cooperation between the people with the information and the person who needs it—YOU.

3. If the circumstances dictate, you may be faced with the suspect who must be detained or arrested even though minimal evidence is available. If this is the case, and the law allows the suspect's arrest, it should be done to prevent an escape and eventual hunt to locate him or her that would threaten the case. This is not to say that you should arrest someone without cause, that is something that should never be done. But you should be prepared to move forward quickly if circumstances so dictate.

4. Find people with information about the case as soon as possible and talk to them right away. This may prove challenging if there are a number of persons who witnessed the event and have information to share with you. Remember, people who witness a crime may not want to share their information with

you. They may not be in a position to talk to the police in front of their friends or associates for fear of retribution. They may have alternate plans that require them to leave the area right away, which makes your time frame for interviewing them unsuitable to their situation. Establishing a way to get in touch with them at a later time with more suitable circumstances is a good thing to do.

5.  Protect any physical evidence by setting up some kind of control or security around the crime scene. Although it may not look like there is any evidence of value during your preliminary check, there may be important clues that a trained evidence technician will be able to locate, process, and testify about in a trial or administrative hearing. It is also important to realize that even though you do not think there is evidence at the scene, every step you take as you walk through the location may be the one that destroys what evidence there was. Remember, once evidence is gone, it can rarely be replaced. Having solid and consistent habits in dealing with evidence will benefit you throughout your career. Whether you are outside at an industrial site or in a bank, taking care of the evidence starts the moment you arrive. Take a good look around and start right off by doing the right thing. This will allow you to establish and maintain a solid chain of custody on the evidence in your cases.

6.  Interviewing a suspect is a good way to get information about what happened, but this is an area greatly determined by the circumstances of the situation, not the least of which is the suspect's attitude and his or her level of cooperation. It may be advisable to wait before doing any suspect interviews until a full assessment of the case at hand can be made. A good rule of thumb is that if in doubt, always get a second opinion from an experienced investigator or a representative from the district attorney's office before proceeding. Knowing all you can about the situation and person you are going to talk to can only help you. There is no substitute for good preparation.

7.  No one can remember everything and now is the time to help your memory by starting to write your investigative notes. Good notes will be invaluable to you when you begin writing the report. Knowing how to take quality notes is a practiced art. Having the right notebook for the job is the first step, but using it successfully is a key to being a good investigator.

8.  Crime scene investigators do not just appear out of thin air. Someone must notify them that they are needed at the crime scene. If you are in charge of the investigation the responsibility to make the notification is yours. Evidence specialists are good at their jobs, so let them help you do yours. This can take some time. The size and complexity of a crime scene dictates how much time will be needed to process a scene. Evidence specialists will document their actions and give you

a report. This part of the investigation should not be rushed. If you are in charge, let them have all the time they need. It will pay big dividends for you in the end.

9.  Get some help. If you need someone with expertise in a certain area, do not hesitate, make the call. You can expect to run into crimes and crime scenes that are new and complex. Get an expert to help you.

10. Write the report.

## HOW IT HAPPENS

In the ideal investigation, information comes to the investigator in a logical and understandable manner. In the real world few if any investigations take place under ideal circumstances. You will probably agree that the investigations of the World Trade Center bombing in New York, and the downing of Pan Am Flight 103 over Lockerbie, Scotland, took place in less than ideal circumstances. They also involved hundreds of investigators and very large crime scenes. This then creates the need to develop information and store it in a systematic way that is accessible by those who need it. It is important for an investigator to recognize that information gathering is the key to solving cases.

The **average investigator** looks at a group of people near a crime scene and says, *"There is a group of people standing around my crime scene."* The **superior investigator** looks at the same group and says, *"There is a group of people standing around my crime scene and each of them knows something about this case. It's my job to get them to tell me what they know and then figure out how to get the ones who don't want to talk to me to do so anyway."*

Some of the things you may be able to find out about the case include who is involved and what their background is, where they are and where they should be, who their associates are, and a lot of information about the crime, especially the **WHO, WHAT, WHERE, WHEN, WHY, and HOW.** These six words, also referred to as the "five W's and the H of investigation," are used by many people to describe what it is they are looking for. However, without some explanation or example, they might remain just words. When good investigators use these words, some of the things they think of are:

### WHO

Who is the reporting party.

Who is involved as the victim.

Who is the suspect.

Who are the suspect's friends and associates.

Who knows what happened.

Who do I need to talk to.

Who is a witness.

Who was with the victim.

## WHAT

What happened.

What was the victim doing.

What was used to commit the crime.

What has happened since I was called.

What do we know so far.

What are the next five things that need to be done.

What time did it happen.

What time was it discovered.

What time was I called.

What was the relationship between the suspect and victim.

## WHERE

Where did it happen.

Where is the victim.

Where is the informant.

Where is the evidence.

Where were the witnesses.

## WHEN

When did it happen.

When was it reported.

When was the victim last seen.

When was the suspect last seen.

## WHY

Why did it happen.

Why was it reported.

Why did witnesses not tell you something.

Why did witnesses tell you certain things.

Why did the crime happen the way it did.

## HOW

How did the event happen.

How did the suspect get there.

How did the suspect get away.

How did the suspect know the victim.

How was the event reported.

How old is the crime.

How much more needs to be done.

There is a lot to be said for the experienced investigator who has seen or done it all a hundred times over, knows all the tricks in the book, and knows when to use the magic that accompanies this knowledge. But much success has also come to those who, although not as experienced, are willing to work hard and learn from the actions of others. One should never rule out luck, or a hunch as a way to turn a dead-end inquiry into a viable and active investigation. This is not downplaying the abilities of the thousands of superior investigators who are very successful at their jobs, but it is to make a point that the uncontrollable facts of daily life may turn out to be a blessing in disguise.

If there were a way to list the characteristics and traits that superior investigators possess and have passed on to one another, it would surely require more than a few pages to do so. There does, however, appear to be some common thread woven through the cloth of the superior investigator that holds together these qualities, which indicate a greater chance for success.

1.  *Superior investigators are ethical.* They inherently know the difference between right and wrong behavior and what is necessary for them to be considered a professional. They play by the rules and do not get involved in the manufacture of false evidence or testimony.
2.  *Superior investigators are aware.* They employ an awareness of their surroundings and are able to quickly evaluate situations and circumstances, as well as people, and are accurate in their assessments. They have a broad knowledge of local and world events, which gives them the ability to place occurrences in perspective and forecast strategies and alternate courses of action with a high probability of success.
3.  *Superior investigators are energized.* The long and tedious hours of fact gathering and analysis do not overwhelm them. The day is never too long to finish some aspect of the case. Even when all seems lost and there are no prospects for success within grasp, the superior investigator can reach within and find the energy and endurance to take the next step, to do one more interview, to read one more supplemental report, or to make one more phone call on the chance it might turn into a workable lead. They give 100 percent all of the time and expect everyone else to do the same.
4.  *Superior investigators think outside the box.* They have the ability to see beyond the next two or three steps in the investigative process and, thus, are able to visualize new or unusual ideas or techniques that may prove to be helpful in solving the case. They are not afraid to try something new or unusual for

**Figure 1-1**

the sake of achieving success in the case. They constantly evaluate the successful ideas and tasks, which they and others routinely perform in a process of continuous improvement.

5. *Superior investigators are determined.* They have the ability to move forward, even in the face of the seemingly unachievable. When others might hang it up and say *"Enough is enough,"* the superior investigators are saying, *"Well, so far I haven't been successful, but I've only tried a few ideas."* Superior investigators never give up. They may prioritize and reprioritize their cases, and put the ones with few or no workable leads on the back burner, but they are always thinking about what they need to do to find the truth.

Just as the job of a good investigator has been described, it is equally important to describe what an investigator's job is not. It is not the investigator's job to prove someone is guilty, or that a particular person did not commit the crime. It is the investigator's job to report **all** information, regardless of the impact the fact pattern may have on the case. An investigation must be unbiased and impartial to be valid. An investigator gains credibility by doing a

**Figure 1-2**

thorough and professional job on all inquiries undertaken. It is not necessary or realistic for an investigator to find a solution for every investigation he or she starts nor is such success necessary to be considered a superior investigator or to have credibility. Superior investigators do all of the little things right the first time, allowing them to start the big things on a positive and timely note. This creates the opportunity for success and allows for the best possible outcome. This in turn allows them to establish credibility, something that is hard to come by, yet easy to lose.

## REVIEW

1.  An investigation is a lawful search for things or people.
2.  The goal of an investigation is to find the truth.
3.  Investigations can start before or after the crime.
4.  Protect the crime scene.
5.  Establish a chain of custody for evidence.
6.  Locate witnesses as soon as possible.
7.  Involve experts in evidence handling.

8. Report all information.
9. Use the who, what, where, when, why, and how to get started.
10. Good investigators are aware, energized, determined, and think outside the box.

## EXERCISES

1. Visit a library or bookstore and review several books about real-life investigations. Select two books of interest to you and while reading them try to determine:
   a. What was the reason for the investigation?
   b. Are there examples of the investigator following or not following the guidelines discussed in Chapter 1?
   c. What superior investigator traits are evident?
   d. In your opinion, which of these traits was most important to the successful completion of the investigation?
2. Match the example of superior traits with the quality
   a. Perceptive
   b. Tireless
   c. Ethical
   d. Creative
   e. Persistent
   _____ Reviews old cases, looking for something new to try.
   _____ Interviews a witness three to four times.
   _____ Tries a new undercover operation.
   _____ Connects names and events in the news to an old case.
   _____ Reprioritizes what needs to be done on a case.
   _____ Willing to work long hours to get the job done.
   _____ Always does the legally and morally correct thing.
   _____ Would never fabricate evidence.
   _____ Uses a sting operation to catch a crook.
   _____ Uses events to forecast likely outcomes and scenarios.
3. Identify a criminal investigation in which the lead detective did not exhibit ethical behavior. What are the consequences for the involved investigator, his or her agency, and the profession?
4. Match the WHO, WHAT, WHERE, WHEN, WHY, and HOW with the example. Answers may vary.
   _____ The television was worth $700.
   _____ Dave was stabbed at 1200 hours.
   _____ The door was kicked in.
   _____ Ron Ho called at 1845 hours reporting gun shots in the area.
   _____ Joe Smith was arrested at 6th and Main.
   _____ Footprints were found outside the broken window.
   _____ The silent alarm was received at 2100 hours.
   _____ Pete Brown was shot in the back of the neck.
   _____ Fava saw a red car drive away.

_____ The suspects took the necklace from the display case.

_____ The knife was marked KCB and booked at Central Property.

_____ Najjar opened the store and saw the broken window.

_____ Just before the shooting, Brown was drinking at the bar.

_____ Bill Smith was lying on the bathroom floor.

_____ Dispatch received the first call at 0630 hours.

_____ The coroner identified the victim as Ron Lincoln.

_____ Detective Colin just left the crime scene.

_____ The body was found under the bridge.

_____ The victim's wife purchased a $300,000 insurance policy last week.

_____ I found a stolen bicycle in the bushes.

5. Read the newspaper for one week and identify the investigations that are reported. Brainstorm the investigative techniques used in the cases. Can you suggest others that might be successful?

6. Working with three to four classmates, construct a bibliography of investigative books. Use this bibliography to develop your knowledge of investigative techniques.

## QUIZ

1. The definition of an investigation is:
   a. An inquiry by a professional
   b. A search for a stolen television
   c. A lawful search for a thing or person
   d. An undercover search of a campaign office

2. The goal of an investigation is to:
   a. Find out what happened
   b. Find out what was taken
   c. Get to the bottom of the matter
   d. Find the truth

3. Generally speaking, the less evidence you have, the better your case will be.
   a. True
   b. False

4. When you begin an investigation, your first consideration should be:
   a. Catching the suspect
   b. Getting there
   c. Protecting the evidence
   d. Locating the best witness
   e. Making an arrest

5. The best person to call to take care of the evidence at a crime scene is:
   a.   A sergeant with 20 years experience
   b.   Your partner
   c.   A crime scene investigator
   d.   A volunteer who has completed an evidence class

6. What do the "five W's and the H of investigation" refer to?
   a.
   b.
   c.
   d.
   e.
   f.

7. Good investigators should report:
   a.   All information
   b.   Only the information that helps show guilt
   c.   About 75 percent of the information they discovered
   d.   Only the information that shows innocence

8. What are the two main ways an investigation can begin?
   a.
   b.

9. Which of the following is not a public record?
   a.   The number of arrests a person has
   b.   How much taxes a property owner pays
   c.   The number of civil actions a person has been a plaintiff in
   d.   The number of convictions a person has

10. Name the qualities of superior investigators
   a.
   b.
   c.
   d.
   e.

# 2 Note Taking

Few people have the ability to remember everything they do, see, or hear. The short-term recall most people have is poor at best and virtually nonexistent at worst. In most cases the longer the time span between the occurrence of an event and the reporting of it, the greater the chance of an incomplete version of the story being told. Further, as time passes it is more likely that inaccurate facts or mistakes will be presented and that something important will be left out of the story. Think back for a few minutes and try to remember what you had for breakfast today, what color shirt you wore yesterday, or who were the first three people you talked to two days ago. You were probably able to remember because they are recent events that you are personally connected to. But try to recall the same information from three weeks ago and you will see it is much more difficult. I think you will agree that the longer it is between something happening and trying to recall it, the more unlikely the latter becomes. Also, consider the example presented by a college class. We all know that there will be a final exam at the end of the semester and quite possibly a midterm exam partway through the course. We also know and recognize the value of having complete and accurate notes in order to study for the exam. Even with this knowledge and the opportunity to take notes and compare them with our classmates, we find ourselves at the end of the semester looking at an exam that contains questions we knew would be on it and have no idea how to answer them.

With regard to investigative report writing we know that the events we are investigating are going to result in questions at a later time. It may be a few hours, days, weeks—or in some cases several months or years in coming—but nonetheless the questions will come. Fortunately, we have the opportunity to do something about this during the investigation, something that will help us when the exam—usually in the form of a trial—may require our testimony. This something is note taking, and it is an essential part of the investigative report writing process. It would be a wonderful world if our memories were so good we never had to refresh them, but that is not the case. You will need notes and you will need to refresh your memory. Fortunately, there are a few basic rules and guidelines that will make the note taking function of report writing much easier to understand and accomplish. Inasmuch as we have discussed why we need to take notes, it may also be helpful to understand some of the basic uses for notes.

## BASIC USES

*Storage.* The old adage "A place for everything and everything in its place" applies here. Both public and private record keeping systems, as efficient as they are, are not perfect enterprises. Over the course of many months and years, documents are misplaced, lost, or destroyed. Whether these events occur by accident or with deliberateness, the end result is that an official document is no longer available. The responsibility for replacing the lost or unavailable documents rests with the person who originally completed the report. Although this may seem unfair to some, it really is the best method to replace whatever is missing. No one else has the knowledge of what the report may contain. In addition, no one else has the original notes that would allow the investigator to talk to the involved parties, and once again find pertinent information. In their most basic form an investigator's notes are the first and last level of record keeping.

*Building block.* A report just does not appear from thin air. Much work and thought goes into writing a thorough and complete investigative report. It is of paramount importance that an investigator has written notes to refer to as he or she completes the report. An analogy can be drawn if you compare the investigative process and report writing tasks to certain parts of a construction project. Investigative note taking and report writing is much like preparing a building site for new construction. A foundation must be completed before the walls can be built. Notes are the foundation and the report that follows is like the building that rises from this first important step. Your notes are the most complete supply of raw material for a report. If you have done your job you should have information from several people involved in the case—descriptions of suspects, property, and evidence and where it was found. In short, you should have what is needed to construct a report and continue working the case.

*Aid to memory.* Investigators with photographic memories may never need to take notes during their investigations because they will be able to remember all of the details. Most investigators do not have photographic memories, however, and do require good notes to remember not only the things to include in the report, but also the unreportable facts. What seems trivial at the earliest stages of an investigation may be very important several months later when an unknown witness or recently discovered evidence surfaces. It seems that a fairly common occurrence in recent years is for convicted criminals to confess to serial crimes that they committed years ago. In order to prove that they did the crimes they offer small details of the crime scenes or victim descriptions that could only be known by someone at the scene. Where is the best place to look for this information? The answer is in the notes taken by the investigation officers. These notes can help officers remember details that no one else would and also help in formulating questions that could eliminate a publicity seeker who is making a false confession. Complete notes are invaluable in these circumstances.

## THE MECHANICS OF NOTE TAKING

The mechanics of note taking refers to the practical application of writing down the initial findings of an investigation in a notebook. It includes the selection of a specific type of notebook and an appropriate writing instrument. The mechanics of note taking are simple to master with practice. It is desirable to take notes in a way that allows anyone who reads them to understand and interpret what they mean. For example, some individuals have handwriting that is difficult to read. This presents a problem if a case is going to involve more than one investigator or requires several people to review and share recorded field notes. There are few rules on the type of handwriting that should be used to record field notes. Whether you print or write in cursive, the end result is what is important. That end result is that you will be able to read and understand the notes when they are needed down the road. Good penmanship is a learnable skill and the adage that "practice makes perfect" certainly applies here. Whenever you are asked to write a lot of information into a small space, there is a tendency to use symbols, codes, or abbreviations to get the job done. Although the use of this type of writing is permissible, it should be avoided as much as possible because different marks and symbols can mean different things to different people. You must also remember that the secret codes or abbreviations you are using today may not mean the same thing to you several months from now. Few things will prove as embarrassing as not being able to read and understand your own notes. Avoid this problem by using clear, complete words whenever possible. The goal of note taking is to record as much information as possible about an investigation in a concise, readable, and understandable manner. Develop good habits and improve on them as you gain experience. It will pay big dividends for you over the length of your career.

There are many types of notebooks that may be used to record field notes. The key here is that the investigator should feel comfortable with the notebook he or she is using, and that it allows for easy writing in a variety of situations. There will be many instances, such as during a surveillance, where the investigator will have to take notes while standing, sitting, walking, or even lying down. It may be necessary to record notes during the day or night, whether it is sunny outside or raining. As you can see, the notebook must be a versatile tool.

In all probability there is no single notebook that meets everyone's needs. Smaller sizes such as a 3″ by 5″ pocket notebook allow for easy storage because it will fit into a shirt or pants pocket. The drawback is that with such small pages it is difficult to record much information. Something like a 6″ by 9″ spiral-bound steno notebook has the advantage of being big enough to write necessary information in it, yet is still manageable enough to easily hold and be stored in a pants pocket or inside the belt line at the small of the back. A full-size pad of paper gives the note taker ample room to write, but may be awkward to hold and store. Another disadvantage is the ease with which pages may become detached from a large tablet of paper.

An investigator should try several sizes and styles of notebooks until the one that best meets his or her needs is found. A suggestion is to hold each type of notebook under consideration and, while standing in an area dark enough to require the use of a flashlight, write enough information to fill five pages. This should give you an idea of the versatility of the notebook and whether or not it will meet your needs. If the notebook passes this test it will likely allow for easy use in any situation or environment an investigator may be involved in, such as in a car, in a house, or in inclement weather. How you organize your field notebook can have an impact on how successful you are in using it. Filling a notebook with information is only part of the battle. You must be able to access the information and retrieve what you need, when you need it.

It is generally recommended that you write on one side of the page. When both front and back sides are used, the notebook can become difficult to handle and manipulate while turning pages and adding information. Use a writing instrument that will not bleed through the page or smear if touched. A good mechanical pencil or ballpoint pen might be a good choice for field officers. Both will give a quality impression on the page and will work in most circumstances. Whatever writing instrument you select, make sure it will stand up to the test of time.

Organizing the layout of a notebook page is another detail that can have a big impact on how useful this tool is for you. One method is to establish a clean break between dates by consistently closing out the bottom of the page by placing some type of ending graphic drawn across the page right after the last entry. Start the next date at the top of a page and begin by writing the date, shift, area, car assigned, and partner's name and serial number if applicable. Create a margin along the left side of the page that can serve as an index to the material. Entries in the margin such as ROBBERY would be followed on the page with the details of the crime or suspect description. Most officers start the day in roll call and receive information about crimes that have happened during the past 24 hours. Indexing these crimes in the left margin will allow you to refer to them quickly when needed. Once you go into the field, add information to your notebook in the chronological order you receive it.

One tendency many new officers have is to write too much on each page. Get in the habit of not crowding too much information together. Leave some space on each line and skip a line or two between different incidents or entries. When the page is set up properly, it will be pleasing to the eye and allow you to glance at it and quickly find what you are looking for. Notebooks are one of the least expensive items you will need in your career. Most agencies supply notebooks to field personnel, but even if you have to buy your own, remember, no cop ever went broke buying a notebook. Leave some room, keep your entries neat, and let the notebook work for you.

Experience is a great teacher in determining what should be included in the note taking process, but there are several items of information that are important to include in every case. Accurate names, addresses, and phone numbers top the list of important information. The importance of recording this type of information accurately cannot be overly stressed. Incorrectly identifying information results in wasted time and an incomplete and sloppy investigation. Other important information includes measurements, drawings, and key words or phrases used by suspects or witnesses. Investigative notes need to be complete enough to help you write an accurate report, but not so in depth that they are as long as the finished product you ultimately produce. Last but not least, the notebook should contain the name, business address, and phone number of the investigator along with the inclusive dates the information in the notebook covers. This identifying information should be clearly printed on the front cover. Remember, this is a business notebook and you are a professional, therefore, the notebook should contain only business related information. Always take care to leave personal information out of the notebook, because it has no place in a professional working document and may be subject to the power of subpoena.

How an investigator takes field notes will depend a great deal on the situation and how comfortable the person being interviewed is about sharing the information they know. Some ability to write quickly and accurately is needed and these are skills that can be improved with practice. As the investigator begins the interview, his or

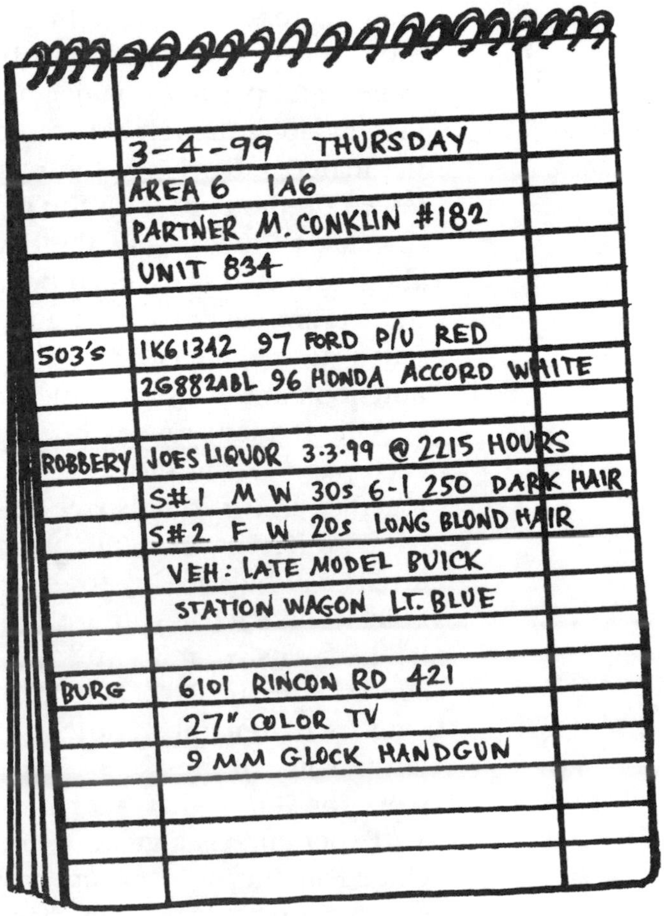

**Figure 2-1**

her manner and demeanor are just as important as the authority or credentials they carry, if not more so. The professional investigator will seek the help of a witness rather than expect it and will convey that attitude in all dealings with those he or she encounters.

A proven method for gaining information in any interview is to start by identifying and introducing yourself to the person you are about to interview and then ask them to tell you a little bit about what happened. If they agree to talk with you and express any cooperation you should let them talk without interruption and without writing anything down. This will give you the opportunity to evaluate their story without being distracted by writing while they are talking. To expedite the interview process ask them to tell you what happened in 30 to 40 words. This is sometimes referred to as the **"thirty word version"** of the story. While they are talking, you may be able to formulate specific questions based on the information they do and do not tell you. Hearing this short version may also help you identify the type of crime or incident involved. Many people will come home, finding that someone has broken into their house and stolen something, and report the event to the police by saying, "I've been robbed." It is easily understood why the victim of such a crime would report it as a robbery when it is really a burglary. However, giving the victim the chance to tell you about the

**Figure 2-2**

crime in their own words gives you the opportunity to formulate questions based on the overview you hear. As well, it gives you the opportunity to determine what happened. Understanding the type of crime is important to the investigator who may need to locate specific evidence or information to establish the corpus, or elements of the crime.

Once the witness has given you the "thirty word version" you should ask specific questions that require specific answers. This will provide you the basic information needed in order to complete the crime or incident report. You should have a thorough knowledge of the format of the crime report forms you are using and should use this format as a guide to formulate and sequence your questions. By doing so you will be able to write the information into your notebook in the same order it will be needed to complete the report. Putting this information into chronological order will save you time and effort preparing the finished product. This is only a suggestion and should never be used as a rigid rule that would limit the amount or quality of information you could gain from someone. Common sense is a great asset for the professional investigator and will almost always allow for a successful investigation.

Once you have completed the task of getting the identifying information written down, ask the witness to tell you what happened and try to write down ideas or major thoughts as they are presented to you. After you have completed the question and answer session with the witness, it is usually a good idea to review the information with them and give them the opportunity to add and change anything that they have misstated to you. Remember, it is your obligation to write an accurate and unbiased report. Giving the victim or witness the chance to alter or correct their story at this time helps to insure that you do your job. It never hurts to conclude the interview by asking if there is anything else they want to add and by giving them a phone number to call you at if they need to speak to you again. The two most important things to remember and practice with regard to note taking for investigative report writing purposes are to **listen first, then write**, and **make sure you can understand your notes.**

## REVIEW

1.   Basic uses of notes are:
      Storage of information
      Building blocks for reports
      An aid to your memory
2.   Notes must be legible and understandable.
3.   Information must be accurate.
4.   The type and style of notebook is an individual matter.
5.   Keep personal and nonbusiness information out of your notes.
6.   Remember to ask for the *thirty word version.*

7. Listen first, then write.
8. Make sure you can understand your notes.

## EXERCISES

1. Review your class notes from any college course or training session you have completed and consider whether or not:
   a. Your notes are an aid to your memory about what took place in the class.
   b. Your notes successfully serve as a storage place for the information you need.
   c. You are able to prepare a one-page report about the class using the notes as a building block for the report.
   d. There is any personal information in the notes.
2. Record a 30-minute news broadcast and watch it while it is recording. Watch it a second time while taking notes and use these notes to prepare a report about the stories that were broadcast. Which principle of note taking will this help you practice?
3. Repeat this exercise with a different tape, but do so while standing up in a dark room using only a flashlight and the light from the television screen to illuminate your notebook. Any difference in the mechanics of note taking?
4. Visit a stationery store and look at several types of notebooks. Make a list of the pros and cons of each notebook you examined. Which type and size do you prefer?
5. Interview a classmate and record enough information so that you can introduce them to the class. Pay particular attention to the correct spelling of names.
6. Watch a television program with a "cops and robbers" theme. Record the information you think is important and then prepare a one-page report from your notes.
7. Review a classmate's notebook. How is it organized? Are the notes legible and understandable?
8. Use a field notebook to sketch a diagram of your classroom and include any audio visual equipment in the room. Be sure to include measurements of the room and the placement of the equipment.
9. Visit the campus library and sketch the first floor and include the measurements of all the important equipment and tables.

## QUIZ

1. Why is it important to have field notes?
   a. Recording all information is required by law
   b. Few people can remember everything
   c. Taking notes gives an officer a way to look professional
   d. Notes offer the proof that something happened

2. List the three basic uses of field notes as discussed in this chapter.
   a.
   b.
   c.
3. Field notes have been compared to the construction of a building site. Explain.
4. Using abbreviations in notes is a good practice to follow whenever possible.
   a. True
   b. False
5. What is meant by the expression *field notes are an aid to an officer's memory?*
6. The preferred type of handwriting for your field notes is:
   a. Short hand
   b. Cursive
   c. Block printing
   d. Anything that is legible and understandable
7. What is the best size of field notebook?
   a. The smaller the better
   b. Anything with removable pages
   c. Clipboard size to get lots of information
   d. Whatever works best for each officer
8. What information should be included at the start of each day's notes?
   a. Your name, service date, and car assignment
   b. The date, your name, area, and car assignment
   c. The date, your partner's name, area, and assignment
   d. The date, your partner's name, area, and car assigned
9. Where should your name and other identifying information be written in a field notebook?
   a. On the cover
   b. On the last page
   c. On both the first and last page
   d. On the first page
10. The basic principle to be followed in taking notes is:
    a. Be polite
    b. Write everything you can
    c. Listen first, then write
    d. Use as many abbreviations as possible

C H A P T E R

# 3 The Rules of Narrative Writing

If you had the ability to place the rules and regulations of all the investigative agencies in this country side by side, you would undoubtedly see that there are thousands of differences in the way they do things. But just as there are differences there are many similarities as well. It is a safe bet that each of these agencies wants its investigators to be courteous, fair, and professional, and to do a good job. It is also a good bet that each of these agencies wants and expects its investigators to document their investigative efforts by writing reports. One of the differences in the way these agencies do business is how reports are completed. It is not uncommon to find a different report form and format from agency to agency, nor is it uncommon to find different forms and formats within a single agency. One reason for this difference is that when investigative reporting is controlled by automation, such as word processing or computer-aided record keeping, these systems sometimes require reports to be in specific formats in order for an interface between the investigator's work and the automated system to occur.

Fortunately, this usually pertains to the face sheet, heading, and "fill in the blanks" type of information. Normally, there are no hard and fast rules about how the narrative portion or body of the report should be written. One method that works in all investigative writings—including crime, arrest, supplemental, incident, evidence, and information reports—is to use a continuous, free-flowing narrative style of writing. With this style no subheadings, side bars, labels, or other text dividers are used. Instead, the investigator follows the **Rules of Narrative Writing** and writes in the first person, past tense, active voice; in chronological order beginning with the date, time, and how he or she got involved; and using short, clear, concise, and concrete words.

## THE 1ST RULE OF NARRATIVE WRITING

### First Person

When writing a report you should write in the first person and refer to yourself as "I." This clearly identifies you as being the person doing the investigation, and it is a clean and simple way of describing who you are. It is one of the shortest words in our language, looks good to those who read your work, and sounds good to those who hear your testimony. As the writer of an investigative report it is always proper to use the personal pronouns "I" and "me" when referring to yourself. For example:

I saw the pry marks on the window sill.

I interviewed Rand.

Colin gave me the pistol.

I found the marijuana.

Najjar and I booked Edwards.

Ho told me he searched Lewis.

## THE 2ND RULE OF NARRATIVE WRITING

### Past Tense

The events you are writing about have already happened and as such are part of history. Inasmuch as they have already taken place it is quite proper to write about them as past events. The verb tense is the part of grammar that tells the time of action, and in standard English there are six tenses. One of these tenses, called the past tense, is used to describe events or actions that have already occurred. Therefore, when writing an investigative report you should use the past tense. For example:

| IF THE VERB IS | THE PAST TENSE WOULD BE |
|---|---|
| To see | I saw Colin. |
| To tell | Fava told me. |
| To say | Ho said he hit Cleworth. |
| To go | We went to 4th and Orange. |
| To find | Conk found Egan inside. |
| To hear | I heard Ito talking. |
| To smell | We smelled the burning paper. |
| To write | I wrote the citation. |

When quoting someone who is telling you what a suspect said, you should still use the past tense for the speaker. For example:

Smith told me the suspect said, "Give me your money."
Brown told me that all Smith said was, "Do it."

By writing in the past tense you will develop consistency, and as a result create a report that is professional and easy to read.

## THE 3RD RULE OF NARRATIVE WRITING

### Active Voice

The active voice is the way of writing that shows who is doing an action, as opposed to the passive voice that shows who is having something done to them. The writer using the active voice tells who is doing a particular action or thing before they tell you who is doing the action or thing. For example:

| ACTIVE | PASSIVE |
|---|---|
| I wrote the report. | The report was written by me. |
| I arrested Brown. | Brown was arrested by me. |
| Smith searched the car. | The car was searched by Smith. |
| I found the gun in the trunk. | The gun was found in the trunk by me. |

One consideration for choosing the active voice over the passive voice is that it is almost always possible to write something in the active voice in fewer words than it would take to write the same thing in the passive voice. When investigators use the passive voice a fairly common problem is that they forget to include who is doing the action, especially with regard to the chain of custody. It is not uncommon to see sentences like:

A bag of marijuana was found in the trunk.
Simmons was arrested.
Cape was read his Miranda rights.
The evidence was booked at the Central Property room.

## THE 4TH RULE OF NARRATIVE WRITING

### Chronological Order

One of the most difficult parts of any writing project is getting started and this is certainly true when it comes to investigative report writing. One popular school of thought is to begin writing at the point of some action being taken, such as what a witness or a victim tells you about the case you are investigating. By doing so you will avoid having to repeat any information that is contained on the face sheet, such as the date of occurrence or the location. While this may save some time and paper in the short run it fails to account for the benefits of starting the report in a way that has long-term and important benefits. One of the keys to successful investigative report writing is to develop consistency, and nowhere is consistency more important from both the reader and content perspective than at the beginning of the narrative.

First, starting a report with the *date, time, and how you got involved* removes the problem of how to start writing. Second, it clearly establishes the reason an investigation was initiated. Granted there may be incidents where information on the face sheet is repeated in the narrative, but these occurrences will be minimal. Although it will be redundant, the benefits of beginning the report in this manner outweigh the costs of doing so.

A third reason for beginning the narrative of all reports in this manner is that not all investigative reports have face sheets. If one were to follow the suggestion and reasoning that no information should be repeated, there would be two rules to follow—one for reports with face sheets and one for reports without. By having one rule to follow and one way of starting the narrative section of a report, the investigator's job will be simplified and chances are that a better work product will be produced.

A fourth reason is that **the big picture** must be considered. There may be instances when several investigators will write reports about a single investigation. Homicides or robberies might involve many investigators, all of whom may be trying to gather information about the same case, but doing so at different locations and times. At some point it may become necessary to incorporate all of the individual reports into an overview for court presentation or a search warrant affidavit. The task of doing either of these is made much easier when the initial sentences of the reports contain all of

**Figure 3-1**

the needed information to place them in the proper order. Time lines or flow charts of the investigation are also easily constructed when the date and time are readily accessible. No one likes having to comb through lengthy written passages to find one or two important facts. Starting all reports in this manner eliminates the problem of how to find the beginning of the investigation.

Although those who argue that it is redundant to begin the narrative portion of a report with the date, time, and how you got involved are sometimes correct in their belief, it is still beneficial to start a report this way. The positive aspects of doing so clearly outweigh the negatives. Some examples of the proper way to begin a narrative section are:

> On 12-14-98 at 0805 hours, I was driving west on 3rd Street when I saw Jones throw a brick through the front window of the house at 4587 3rd Street.
> On 7-7-99 at about 1910 hours, I received a radio call to investigate a robbery at the Big Burger Drive Inn, 123 Main Street.
> On 3-6-99 at 1245 hours, Sergeant Thompson told me to investigate a reported child endangerment at 6703 Ocean, Apartment 3.
> On 12-22-98 at 1500 hours, I was driving through the south parking lot of the Applewood Shopping Center and saw Brennon run out of Play the Records.

Using specifics to begin the narrative is preferable to the method of using overworked phrases, such as:

> On the above date and time . . .
> Witness Wilson arrived home and . . .
> Victim states that . . .

It is important to remember that consistency is a key to success and beginning all reports in a consistent manner provides this important feature.

---

## THE 5TH RULE OF NARRATIVE WRITING

### *Short, Clear, Concise, and Concrete Words*

All writers have the option of which word they will use to describe what they want their readers to understand. As an investigative report writer you should always strive to get the most out of the words you choose for the reports you write. If the ends of the word spectrum are concrete at one end and abstract at the other, then abstract words are those that have no specific meaning and are open to interpretation. Concrete words are those that have a clear meaning and little or no misinterpretation of their use.

Abstract words are those that could have multiple applications in the context of a sentence depending on the viewpoint of the reader. For example, the word CONTACTED is widely used in investigative reports and has several potential meanings when used in the sentence:

I contacted Smith.

Does the word mean that the investigator touched Smith physically as in the sense that professional football is a contact sport? Does the word imply that the investigator knows Smith is there in the sense that a radar operator has contact with an airplane as it enters his or her control zone? Does it mean the investigator spoke to Smith?

Concrete words are the opposite of abstract words and are generally better suited for investigative report writing. Words like TALKED, SAW, FOUND, SEARCHED, and DROVE are examples of concrete words that investigators should use whenever possible. When you use concrete words you will be able to more clearly describe what it is you have done or discovered in your investigation. For investigative report writing purposes, writers should always strive to write at the lowest level of abstraction, which means they should use concrete words whenever possible. Other examples of commonly used abstract words and their concrete counterparts are:

abstract: I detected the odor of burning marijuana.

concrete: I smelled burning marijuana.

abstract: We proceeded to the jail.

concrete: We drove to the jail.

abstract: I observed Murphy driving north on Main.

concrete: I saw Murphy driving north on Main.

abstract: Rogers indicated he stole the money.

concrete: Rogers said he stole the money.

abstract: I found the weapon under the couch.

concrete: I found the revolver under the couch.

The five basic rules of narrative writing can be applied to any report writing situation and will allow for the successful completion of any investigative report. They were developed to help investigators achieve consistency in their writing and allow them to complete the writing task as quickly as possible.

While considering all the things that can be done to improve writing style, it becomes apparent that there are many areas that provide an opportunity for improvement. In addition to the five basic rules of narrative writing, there are other things the investigator can do to enhance the report and improve its professional quality.

## Minimize the Use of Abbreviations

One of these considerations is the use of abbreviations—or more appropriately—not using abbreviations. Many investigators believe they can save time by using abbreviations because it will reduce the number and length of words in their report, and accordingly, the report will be shorter. However, the end result of such thinking is that the report does end up shorter than if the words were spelled out in full, but along with a shorter report there is the fact that many people who read the report will find it confusing. Although it is not realistic to avoid using abbreviations in all situations, they should be avoided for a complete spelling whenever possible. Instead of using an abbreviation try to find a complete word that means the same thing. An example of this is the standard law enforcement and investigative word APPROXIMATELY. Many investigators would not hesitate to use this word in a variety of situations, nor would they hesitate to abbreviate it as APPROX. To abbreviate it in this manner requires seven characters to be made, including the period at the end. If the same investigator used the word ABOUT in place of APPROXIMATELY, less character marks would be made and the same meaning would be shown. As we know, there are situations in which abbreviations are appropriate and it makes sense to use them. In these cases *use a standard abbreviation that is approved by your agency or company.* Use the complete word if there is no commonly used abbreviation. Do not create an abbreviation, but if you must use one be careful. No one will disagree that a short report takes less time to complete than a long one, but this does not mean that you should sacrifice clear meaning for the sake of brevity.

## Use Last Names without Titles

Another problem for investigative report writers is how to refer to the people they write about. When and how to use names and titles is a dilemma not easily solved. One of the oldest traditions in investigative report writing is to refer to the people involved in the investigation by titles such as "the victim," "the reporting party," or as "witness Fava." Instead of using these titles, use the person's last name. For example, "Brown said," or "I was talking to Kelly," or "I saw Snyder using the shredder." There is nothing wrong with calling people by their names. In fact, it will make the report easier to understand and be more readable. You should also avoid using the titles of Mr., Mrs., and Ms. Just use the person's last name and let it go at that. By eliminating tiles prefixed to a person's surname you will save a lot of writing and keep the report looking a lot neater. Remember, if the report is neat and easy to read, it is more useful than one that is cluttered and sloppy.

The titles of law enforcement officers may also be avoided after the first time the person is introduced in the report. If a person is listed on the face sheet, or some other type of cover page, it is proper to begin referring to him or her by their last name only. There is no need to use the person's title.

There may be instances when you talk to a person who has such minimal or no involvement in a case that they should not be listed on a face sheet, or the report you are preparing has no face sheet. The names of these persons should be included in the narrative and as completely identified as possible the first time they are mentioned. From that point on it is proper to refer to them by their last name only. For example:

> I spoke to John Brown, 1631 Ninth St., Los Angles, CA 90085, (213) 555-1212, and he told me he did not know Johnson.

Should it be necessary to refer to this person at a later time in the report it would be appropriate to use the last name only. For example:

> After speaking to Colin, I called Brown at his home and he told me he had not been at the hospital the night of the theft.

Sometimes there will be more than one person involved in the investigation with the same last name. In these cases use their full first and last name, not the first initial and last name. You are trying to create a document that can be read in the same way a conversation is heard. Not many people will tell you that they saw J. Brown at the market yesterday; they are much more likely to tell you they saw John Brown at the market. Although you should not write the way people speak, you should try to write in a conversational style. Do not use their first names only. This is not professional.

When a situation arises where you have a suspect and you do not know his or her name until you are well into the investigation, there is always the question of how to refer to the person in the report. A good example of this occurring is in a driving under the influence arrest. Normally, these cases begin when a officer stops a car being driven erratically, and has a conversation with the driver. In this example the driver, David Thomas, is arrested by Officer Joe Fava for the crime of driving while under the influence of alcohol. After booking Thomas, Fava begins writing a report. Fava would be correct in beginning the report with the date, time, and how he got involved and in so writing, using Thomas's name. For example:

> On 6-13-99 at 2145 hours I saw Thomas driving a red Pinto, license 123 ASD, north on Main Street from 5th.

In most cases it is not necessary to go into great detail about how you identified a person. If the identity of the suspect was crucial to the commission of the crime a corresponding level of information would be needed to establish identity. This, however, is the exception rather than the rule in most cases. Such phrases as, "I saw the suspect, later identified as Beer, driving in an erratic manner" are not needed. It is proper to begin by writing, "I saw Beer driving in an erratic Manner." Another example of this is the case where police officer George Najjar gets anonymous information that a person is going to rob the Brand X liquor store sometime after 1:00 PM the following day. Najjar begins a surveillance of the store and several hours after the investigation begins sees a person whose name is James Dalton enter the store and rob it. The information was received at 7:00 PM on July 3, 1998. Following the rules of narrative writing the report would begin:

> On 7-3-98 at 1900 hours, I received an anonymous telephone call in which the caller told me that the Brand X liquor store would be robbed sometime after 1300 hours the next day. On 7-4-98 about 1100 hours, I began watching the store and at 1400 hours I saw Dalton walk into the store and . . . .

In this example the officer did not know Dalton's name when he first saw him go into the store, but it is acceptable to use his name in the report. The rule of thumb is that if you know the person's name before you start writing the report, use it from the beginning. It makes for a more readable report.

## Radio Code and Investigative Jargon

Nearly every occupation or profession has words, expressions, and phrases particular to it that are clear to those in the field, yet have no meaning to those outside the occupation. The investigative field is no exception. There are hundreds of investigative jargon phrases that are used daily that give meaning and direction to the activities of investigative personnel, but which do not clearly convey the same message to those outside the field. It is because of this unfamiliar ground and the multiple meanings of the words and phrases that make their use in investigative reports undesirable and improper. Remember that you are writing for a wide audience and that most people in the audience probably have no understanding of the vernacular of your specific field. As such, these expressions and phrases should not be used.

The use of radio code is another area full of opportunities for confusion and miscommunication. Radio code is a way to communicate over public air waves in a shortened way that reduces the air time needed to get a certain message transmitted. Radio code is appropriate when talking on the radio or when involved in a conversation where others might be present who should not hear what is being said and there is no other way to safeguard the privacy of the conversation. An example of this would be if an overheard remark would jeopardize an investigation or reveal and compromise what was about to happen in an investigation where timing was crucial to its success. In theory all investigative agencies that adopt certain radio codes also adopt the same translations. It is important to realize that all investigative agencies do no use the same radio code. The reality of the situation is that law enforcement agencies within 25 miles of each other using the same radio code designations many times have different interpretations of those designators. Officers and investigators from these agencies might find themselves hearing a familiar sound but interpret the message in a different manner. If this is the case with professional police officers and investigators, one can only imagine what effect the use of radio code in an investigative report might have on those who are unfamiliar with radio code and who read the report. Investigative reports can be confusing enough without the added burden of having to decipher unnecessary radio codes. Remember, your job is to report what happened and what you did, so that everyone will understand what is going on.

## When to Use Direct or Indirect Quotes

Another area of concern for the investigative report writer is that of quotations. When to quote and how to write it in the report are substantial issues to resolve. The majority of investigative report writing involves the reporting of information gained through one of the senses. There are few people who have the ability to listen to someone and recreate the entire conversation at a later time. More likely, a witness will remember the important things, from their perspective, that someone said or told them. They will tell these important details to the investigator whose job it is to write about them.

The two types of quotes used in investigative report writing are direct and indirect. *Direct quotes should be used when it is important to know exactly what was said.* Many suspects use the same words or phrases when they commit crimes. This then becomes part of their modus operandi and, as such, it is important to quote what was said. It is also important to quote what a suspect says if they are admitting guilt during an interview. There is no better way to describe what a suspect said or did in the commission of a crime than through the use of his or her own words. Accordingly, there are two times when it is very important to quote directly:

1. When a witness tells you what a suspect said during the commission of a crime.
2. When a suspect admits guilt.

Otherwise, it is appropriate to indirectly quote or paraphrase what someone says. The reason it is not crucial to quote everyone all of the time is that if it becomes necessary for the witness to testify in court or at an administrative hearing about what he or she saw or heard, they will be the one to do it. As the investigator your role will be to act as an assistant to the prosecutor or person presenting the case. Generally, you will not be allowed to testify about what someone told you so the need to quote verbatim is not present. Quoting someone indirectly is giving the essence of what they told you. The general idea of what someone said is what you are after, not the exact words. You should remember, however, that if you have the opportunity to quote what a suspect is telling you and you do not, you are giving away the advantage to the suspect.

## REVIEW

1. Start the narrative of all reports the same way.
2. The five rules of narrative writing are:
   Write in the first person.
   Use the past tense.
   Use the active voice.
   Start with the date, time, and how you got involved.
   Use short, simple, concise, and concrete words.
3. Avoid using abbreviations.
4. Refer to people by their last names.
5. Avoid using titles like Mr., Mrs., and Ms.
6. Keep radio code and jargon out of the report.
7. Use direct quotes only when needed.

## EXERCISES

1. Visit a library and review several publications dealing with trends in writing. Is the continuous, free-flowing, narrative style of writing as described in this chapter discussed in any of your readings? Are there any differences between the rules discussed in your outside readings and those in this chapter? If so, list and discuss them.
2. Locate an investigative report written in the passive voice and re-write it using the active voice. Were you able to say the same thing in fewer words?
3. Locate an investigative report that is not written according to the rules of narrative writing. What differences do you see? How can these differences be corrected?
4. What is the difference between the first- and third-person style of writing? What are the advantages of using the first-person style in investigative reports?
5. First-person exercise. Change the sentences to the first person style. For the purposes of the exercise, you are Officer Fava and your partner is Officer Ho.
   a. U/S officer found the pistol in the street.
   b. Sergeant Najjar gave the evidence to this officer.
   c. Undersigned Officer Fava interviewed Brown.
   d. Officer Fava and Officer Ho searched the car.
   e. U/S officers were assigned to guard the scene.
   f. Cleworth gave the video to Officer Fava.
   g. Officer Donald and Officer Tatum showed this officer where they found the stolen diamond.

    h.   It was Investigating Officer Fava's intent to question Jackson as soon as possible.

    i.   Officer Fava's and Officer Ho's report is in the watch commander's office.

    j.   Officer Ho and I investigated the theft.

6.  Past-tense exercise. Change the following examples into the past tense.

    a.   I am going to the hospital.

    b.   We are finding the evidence.

    c.   Booth analyzes the evidence.

    d.   Last Tuesday, I was watching the intersection.

    e.   After briefing we proceed to the central jail.

    f.   At 1130 hours we (eat, ate) lunch.

    g.   You (see, saw) the marijuana plants.

    h.   We (arrest, arrested) Roberts at 123 Main.

    i.   I (book, booked) the knife at Central Property.

    j.   We (drive, drove) to the hospital.

7.  Active-voice exercise. Rewrite the sentences in the active voice.

    a.   The call was answered by Rogers.

    b.   Smith was read his Miranda rights by me.

    c.   The car was searched by Stone.

    d.   Ten area cars were wanted by the Chief.

    e.   The gun was found by Thompson.

    f.   The report was written by the lieutenant.

    g.   Cleveland was booked by Winters.

    h.   The longest speech was given by the Mayor.

    i.   The window was broken by Watson.

    j.   The evidence was examined by Wilson.

8.  Chronological order exercise. If not shown, use today's date and military time to establish the proper starting sentence for a narrative report.

    a.   It was 3:00 PM when I was dispatched to the Mercy Hospital emergency room.

    b.   On Sunday, the 17th of January, approximately 2:00 AM, I found a person, later identified as Sproul, sitting on the steps of the East Branch Library.

    c.   Roberts hailed me on March 15, 1999 at about 8:15 AM while I was driving by the gas station.

    d.   Unit 12, which I was driving, was radioed to go to Main and Temple at 6:30 PM to see a man about a theft.

    e.   At 1605 hours on April 6, 1999, I was dispatched to the Big Burger Drive Inn regarding a disturbance.

    f.   On 7-4-99 at about 2215 hours, we saw Potts throw a lighted flare into a dumpster at the card board recycling plant, 1641 West Street.

    g.   On 24 March 1999 at 2:05 AM I heard a traffic collision at Washington and Dysart.

    h.   On 12-24-99 at 2100 hours I was dispatched to Highway and Noble Drive regarding a theft of Christmas trees in progress.

     i.     On Tuesday, February 16, I saw Donaldson driving west on Elm approaching Pacific.

     j.     On 5-26-99 at approximately 1305 hours, Smith assigned me to investigate a theft at the County Credit Union.

9.    Short, clear, concise, and concrete word exercise. Suggest a better word to be used in place of each abstract word shown.

    a.   proceeded
    b.   contacted
    c.   detected
    d.   advised
    e.   indicated
    f.   weapon
    g.   gun
    h.   demonstrated
    i.   stated
    j.   observed

10.   Why should the use of jargon be avoided in investigative reports? Give some examples.

## QUIZ

1.    With regard to investigative report writing, what does narrative writing mean?

2.    When should abbreviations be used in investigative report writing?

3.    List the five rules of narrative writing.

    a.

    b.

    c.

    d.

    e.

4.    How should the narrative section of a report begin?

5.    What are three reasons for beginning a narrative in this manner?

6.    Describe the First Person style of writing.

7.    Why should reports be written in the past tense?

8.    Describe the active voice.

9.    What is the rule of thumb regarding the use of radio code in the narrative of an investigative report?

10.   What are the two types of quotes used in investigative reports?

    a.   Accurate and exact
    b.   Exact and indirect
    c.   Gist and accurate
    d.   Direct and accurate
    e.   Direct and indirect

# 4

# Describing Persons and Property

**How to distinguish between**
>   **Victims**
>   **Reporting parties**
>   **Suspects**
>   **Others**

**When to list someone as a suspect**

**How to write a good description of a suspect**

**How to describe property and determine a value**

**The average person test**

**The chain of custody**

**Completing evidence reports**

Crime fighters on the silver screen, and for that matter the big screen television sets in our living rooms, always seem to make the art of investigation look easy, including the documentation of suspect descriptions and those of the property that is damaged or stolen. Crime fighters in the trenches know it is not that simple.

## DESCRIBING PEOPLE IN A REPORT

Report face sheets have spaces for a variety of people who have a part in the report process. These people include victims, reporting parties, suspects, witnesses, and in some cases a large uninvolved group known as OTHERS. Who are these people and how are they defined? When is it appropriate to list someone as a suspect and when is it not? Who is a witness and when does a person qualify as an OTHER?

### Categorizing People for a Report

#### Victims

The definition of a victim is someone who has been hurt, or who may have had some of their property damaged or stolen. They are people who have been the recipient of some wrongful act or deed. Victims may be persons or entities such as school districts, businesses, or corporations. Although the victim is almost always present when an investigator is looking into a crime, it is not necessary for the victim to be present in order to document the event in a crime report. Neither is it necessary to have the victim present to convict someone of a crime. The victim will almost always identify him or herself when the investigator arrives and begins the preliminary work of finding out what happened because many times the victim is the person who reported the crime to the investigative agency. In these cases the victim and the reporting party are one and the same.

Some crimes are categorized based on the characteristics of the victim and of the crime. For example, some burglars only attack residences while others specialize in commercial establishments or businesses. Other burglars operate only at night and will commit crimes when people are inside the location they are burglarizing. The three major categories of burglary are residential, commercial, and vehicle. The distinction is determined by the type of target the burglar selects. Persons who commit robberies often specialize in a certain type. For example, some only attack persons who are on the street, while others rob businesses or residences. Therefore, from an investigative standpoint the categories of robbery are street, residential, and commercial.

A rule of thumb for determining whether the victim of a crime is a person or a business is to determine if the damaged or stolen property belonged to an individual or to a business. This is an area where department guidelines and the investigator's experience will help define the victim.

It may be possible to have both commercial and personal victims involved in the same crime. Consider the robber who enters a bank and during the ensuing robbery takes money belonging to the bank, as well as the personal money belonging to the customers who were inside the bank. If this happened there would be several victims, including the bank and each person robbed. In all probability an agency's guidelines will determine how the crime is classified. In accordance with these guidelines a determination would also be made as to who would investigate it.

### Reporting Parties

The person who reports the crime to an investigative agency is called the "reporting party" or "person reporting." This person may have no direct knowledge of the crime or know anything about it other than they were in a position to summon help. With the increased popularity of cellular phones it is becoming more and more common to have people drive by something they think is a crime and anonymously report it to a law enforcement agency. The uninvolved person who is walking by a store and hears a person inside yelling for someone to call the police may do so even without knowing if a crime has been committed. It is also possible for the reporting party to be involved as a victim or as a witness and be within the scope of both of these two categories on the crime report face sheet. When this occurs it should be noted because one of the purposes of the crime report face sheet is to organize information.

### Suspects

The decision to list someone as a suspect in a crime report is often determined by agency guidelines that correspond to an internal auditing and investigation procedure. Generally, this auditing and investigative procedure dictates that suspects be divided into two categories—felons and misdemeanants. With regard to felony crime investigations you should be aware that in many jurisdictions a felony suspect may be arrested on suspicion that he or she committed a crime. As such, listing a person as a suspect in a felony crime report may be reason enough to arrest him or her if found by a police officer. Therefore, a person should be listed by name on a felony crime report only if there is sufficient probable cause articulated in the report to justify an arrest. If insufficient probable cause does not exist to justify an arrest they should not be listed as a NAMED SUSPECT. It would be appropriate in this case to write SEE NARRATIVE in the suspect area of the face sheet and put the suspect information in the text of the report. This will allow the information to get to the follow-up investigators while protecting the person's rights. (See figures 4-1 and 4-2.)

**EL SEGUNDO POLICE DEPARTMENT** — SUSPECT REPORT

PAGE_____ OF _____

CASE NO

**CRIME 1**

| CODE SECTION | CRIME | CLASSIFICATION | REFER OTHER REPORTS |
| --- | --- | --- | --- |

| LOCATION (Be Specific) | RD. | DATE | TIME | SUPPL. ☐ | INCIDENT NO. |
| --- | --- | --- | --- | --- | --- |

**SUSP. VEH 2**

| LICENSE # | STATE | VEH. YR | MAKE | MODEL | BODY STYLE |
| --- | --- | --- | --- | --- | --- |

BODY STYLE: ☐ 0 UNK ☐ 2 4-DR ☐ 4 P/U ☐ 6 VAN ☐ 8 RV ☐ 10 OTHER
☐ 1 2-DR ☐ 3 CONV ☐ 5 TRUCK ☐ 7 S/W ☐ 9 M/C

| COLOR/COLOR | OTHER CHARACTERISTICS (i.e. T/C Damage, Unique Marks or Paint, etc.) | DISPOSITION OF VEH. |
| --- | --- | --- |

REGISTERED OWNER

**SUSPECT 3**

| SUSP. # | NAME (First, Last, Middle) | SEX | RACE |
| --- | --- | --- | --- |

SEX: ☐ 1. M ☐ 2. F
RACE: ☐ 0 UNK ☐ 2 HISP ☐ 4 IND ☐ 6 JAP ☐ 8 OTH_____
☐ 1 WHT ☐ 3 BLK ☐ 5 CHI ☐ 7 FIL ☐ 9 P.ISL.

AKA | D.O.B. | AGE | HT. | WT. | BUILD

BUILD: ☐ 1 THIN ☐ 3 HEAVY ☐ 0 UNK ☐ 2 MED ☐ 4 MUSCLR

HAIR: ☐ 0 UNK ☐ 2 BLK ☐ 4 RED ☐ 6 S/P ☐ 8 OTHER
☐ 1 BRN ☐ 3 BLN ☐ 5 GRAY ☐ 7 WHT

EYES: ☐ 0 UNK ☐ 2 BLK ☐ 4 GRN ☐ 6 GRAY
☐ 1 BRN ☐ 3 BLU ☐ 5 HAZEL ☐ 7 OTHER ____

D.L. #

| RES. ADDRESS | RD | ZIP CODE | RES. PHONE # ( ) | S.S # |
| --- | --- | --- | --- | --- |

| BUS. ADDRESS | RD | ZIP CODE | BUS. PHONE # ( ) | OCCUPATION |
| --- | --- | --- | --- | --- |

| CLOTHING | ARRESTED ☐ 1 YES ☐ 2 NO | STATUS ☐ 1 DRIVER ☐ 3 PED. ☐ 2 PASS | GANG AFFILIATION: HOW KNOWN: ☐ 1 KNOWN ☐ 2 SUSPECTED |
| --- | --- | --- | --- |

| AMT. OF HAIR 4 | HAIR STYLE 8 | COMPLEXION 10 | TATTOOS/SCARS 13 | DISTING. MARKS 14 | WEAPON(S) 17 |
| --- | --- | --- | --- | --- | --- |
| ☐ 0 UNKNOWN Q21 | ☐ 0 UNKNOWN Q25 | ☐ 0 UNKNOWN Q27 | ☐ 0 UNKNOWN | ☐ 0 NONE Q30 | ☐ 0 UNKNOWN ☐ 0 NONE Q33 |
| ☐ 1 THICK | ☐ 1 LONG | ☐ 1 CLEAR | ☐ 1 FACE _____ | | ☐ 1 CLUB _____ |
| ☐ 2 THIN | ☐ 2 SHORT | ☐ 2 ACNE | ☐ 2 TEETH _____ | | ☐ 2 HAND GUN _____ |
| ☐ 3 RECEDING | ☐ 3 COLLAR | ☐ 3 POCKED | ☐ 3 NECK _____ | | ☐ 3 OTHER UNK GUN _____ |
| ☐ 4 BALD | ☐ 4 MILITARY | ☐ 4 FRECKLED | ☐ 4 R/ARM _____ | | ☐ 4 RIFLE _____ |
| ☐ 5 OTHER_____ | ☐ 5 CREW CUT | ☐ 5 WEATHERED | ☐ 5 L/ARM _____ | | ☐ 5 SHOT GUN _____ |
| **TYPE OF HAIR 5** | ☐ 6 RIGHT PART | ☐ 6 ALBINO | ☐ 6 R/HAND _____ | | ☐ 6 TOY GUN _____ |
| ☐ 0 UNKNOWN Q22 | ☐ 7 LEFT PART | ☐ 7 OTHER_____ | ☐ 7 L/HAND _____ | | ☐ 7 SIMULATED _____ |
| ☐ 1 STRAIGHT | ☐ 8 CENTER PART | **GLASSES 11** | ☐ 8 R/LEG _____ | | ☐ 8 POCKET KNIFE _____ |
| ☐ 2 CURLY | ☐ 9 STRAIGHT BACK | ☐ 0 UNKNOWN Q28 | ☐ 9 L/LEG _____ | | ☐ 9 BUTCHER KNIFE _____ |
| ☐ 3 WAVY | ☐ 10 PONY TAIL | ☐ 0 NONE | ☐ 10 R/SHOULDER _____ | | ☐ 10 OTH. CUT/STAB INST _____ |
| ☐ 4 FINE | ☐ 11 AFRO/NATURAL | ☐ 1 YES (No Descrip.) | ☐ 11 L/SHOULDER _____ | | ☐ 11 HANDS/FEET _____ |
| ☐ 5 COARSE | ☐ 12 PROCESSED | ☐ 2 REG GLASSES | ☐ 12 FRONT TORSO _____ | | ☐ 12 BODILY FORCE _____ |
| ☐ 6 WIRY | ☐ 13 TEASED | ☐ 3 SUN GLASSES | ☐ 13 BACK TORSO _____ | | ☐ 13 STRANGULATION _____ |
| ☐ 7 WIG | ☐ 14 OTHER_____ | ☐ 4 WIRE FRAME | ☐ 14 OTHER _____ | | ☐ 14 TIRE IRON _____ |
| ☐ 8 OTHER_____ | **FACIAL HAIR 9** | ☐ 5 PLASTIC FRAME | _____ | | ☐ 15 OTHER _____ |
| **HAIR CONDITION 6** | ☐ 0 UNKNOWN Q26 | ☐ Color_____ | | | **WEAPON FEATURE 18** |
| ☐ 0 UNKNOWN Q23 | ☐ 0 N/A | ☐ 6 OTHER_____ | **UNIQUE CLTHNG 15** | **WEAPON IN 16** | ☐ 0 UNKNOWN ☐ 0 NONE Q34 |
| ☐ 1 CLEAN | ☐ 1 CLN SHAVEN | **VOICE 12** | ☐ 0 UNK ☐ 0 NONE | ☐ 0 UNKNOWN Q32 | ☐ 1 ALTERED STOCK _____ |
| ☐ 2 DIRTY | ☐ 2 MOUSTACHE | ☐ 0 UNKNOWN Q29 | ☐ 1 CAP/HAT Q31 | ☐ 0 N/A | ☐ 2 SAWED OFF _____ |
| ☐ 3 GREASY | ☐ 3 FULL BEARD | ☐ 0 N/A | | ☐ 1 BAG/BRIEFCASE | ☐ 3 AUTOMATIC _____ |
| ☐ 4 MATTED | ☐ 4 GOATEE | ☐ 1 LISP | ☐ 2 GLOVES | ☐ 2 NEWSPAPER | ☐ 4 BOLT ACTION _____ |
| ☐ 5 ODOR | ☐ 5 FUMANCHU | ☐ 2 SLURRED | | ☐ 3 POCKET | ☐ 5 PUMP _____ |
| ☐ 6 OTHER_____ | ☐ 6 LOWER LIP | ☐ 3 STUTTER | ☐ 3 SKI MASK | ☐ 4 SHOULDER HOLSTER | ☐ 6 REVOLVER _____ |
| **R/L HANDED 7** | ☐ 7 SIDE BURNS | ☐ 4 ACCENT | | | ☐ 7 BLUE STEEL _____ |
| ☐ 0 UNKNOWN Q24 | ☐ 8 FUZZ | Describe_____ | ☐ 4 STOCKING MASK | ☐ 5 WAISTBAND | ☐ 8 CHROME/NICKEL _____ |
| ☐ 1 RIGHT | ☐ 9 UNSHAVEN | _____ | | ☐ 6 OTHER_____ | ☐ 9 DOUBLE BARREL _____ |
| ☐ 2 LEFT | ☐ 10 OTHER_____ | ☐ 5 OTHER_____ | ☐ 5 OTHER_____ | | ☐ 10 SINGLE BARREL _____ |
| | | | | | ☐ 11 OTHER _____ |

| REPORTING OFFICER | ID# | DATE | REVIEWED BY | ID# | DATE |
| --- | --- | --- | --- | --- | --- |

COPIES: ☐ CHIEF ☐ CII ☐ PATROL ☐ DB ☐ OTHER AGENCY     ROUTED BY     ENTERED BY
TO: ☐ ☐ CAU ☐ ABC (2 copies) ☐ DA

ESPD Form #200 (Rev 5/97)

**Figure 4-1** (Courtesy of the El Segundo Police Department)

| CASE NO. |
|---|
| PAGE _____ |

**3 SUSPECT**

| SUSP. # | NAME (First, Last, Middle) | SEX<br>☐ 1 M<br>☐ 2 F | RACE | ☐ 0 UNK  ☐ 2 HISP  ☐ 4 IND  ☐ 6 JAP  ☐ 8 OTH_____<br>☐ 1 WHT  ☐ 3 BLK  ☐ 5 CHI  ☐ 7 FIL  ☐ 9 P.ISL. |
|---|---|---|---|---|

| AKA | D.O.B. | AGE | HT. | WT. | BUILD  ☐ 0 UNK  ☐ 1 THIN  ☐ 2 MED  ☐ 3 HEAVY  ☐ 4 MUSCLR |
|---|---|---|---|---|---|

| HAIR  ☐ 0 UNK  ☐ 2 BLK  ☐ 4 RED  ☐ 6 S/P  ☐ 8 OTHER_____<br>☐ 1 BRN  ☐ 3 BLN  ☐ 5 GRAY  ☐ 7 WHT_____ | EYES  ☐ 0 UNK  ☐ 2 BLK  ☐ 4 GRN  ☐ 6 GRAY<br>☐ 1 BRN  ☐ 3 BLU  ☐ 5 HAZEL  ☐ 7 OTHER_____ | D.L. # |
|---|---|---|

| RES. ADDRESS | RD | ZIP CODE | RES. PHONE #<br>(   ) | S.S # |
|---|---|---|---|---|

| BUS. ADDRESS (School) | RD | ZIP CODE | BUS. PHONE #<br>(   ) | OCCUPATION |
|---|---|---|---|---|

| CLOTHING | ARRESTED  ☐ 1 YES  ☐ 2 NO | STATUS  ☐ 1 DRIVER  ☐ 3 PED.  ☐ 2 PASS | GANG AFFILIATION:<br>HOW KNOWN: | ☐ 1 KNOWN  ☐ 2 SUSPECTED |
|---|---|---|---|---|

| AMT. OF HAIR 4 | HAIR STYLE 8 | COMPLEXION 10 | TATTOOS/SCARS 13 | DISTING. MARKS 14 | WEAPON(S) 17 |
|---|---|---|---|---|---|
| ☐ 0 UNKNOWN | ☐ 0 UNKNOWN | ☐ 0 UNKNOWN | ☐ 0 UNKNOWN | ☐ 0 NONE | ☐ 0 UNKNOWN  ☐ 0 NONE |
| ☐ 1 THICK | ☐ 1 LONG | ☐ 1 CLEAR | ☐ 1 FACE | _____ | ☐ 1 CLUB _____ |
| ☐ 2 THIN | ☐ 2 SHORT | ☐ 2 ACNE | ☐ 2 TEETH | _____ | ☐ 2 HAND GUN _____ |
| ☐ 3 RECEDING | ☐ 3 COLLAR | ☐ 3 POCKED | ☐ 3 NECK | _____ | ☐ 3 OTHER UNK GUN _____ |
| ☐ 4 BALD | ☐ 4 MILITARY | ☐ 4 FRECKLED | ☐ 4 R/ARM | _____ | ☐ 4 RIFLE _____ |
| ☐ 5 OTHER_____ | ☐ 5 CREW CUT | ☐ 5 WEATHERED | ☐ 5 L/ARM | _____ | ☐ 5 SHOT GUN _____ |
| **TYPE OF HAIR 5** | ☐ 6 RIGHT PART | ☐ 6 ALBINO | ☐ 6 R/HAND | _____ | ☐ 6 TOY GUN _____ |
| ☐ 0 UNKNOWN | ☐ 7 LEFT PART | ☐ 7 OTHER_____ | ☐ 7 L/HAND | _____ | ☐ 7 SIMULATED _____ |
| ☐ 1 STRAIGHT | ☐ 8 CENTER PART | **GLASSES 11** | ☐ 8 R/LEG | _____ | ☐ 8 POCKET KNIFE _____ |
| ☐ 2 CURLY | ☐ 9 STRAIGHT BACK | ☐ 0 UNKNOWN | ☐ 9 L/LEG | _____ | ☐ 9 BUTCHER KNIFE _____ |
| ☐ 3 WAVY | ☐ 10 PONY TAIL | ☐ 0 NONE | ☐ 10 R/SHOULDER | _____ | ☐ 10 OTH. CUT/STAB INST _____ |
| ☐ 4 FINE | ☐ 11 AFRO/NATURAL | ☐ 1 YES (No Descrip.) | ☐ 11 L/SHOULDER | _____ | ☐ 11 HANDS/FEET _____ |
| ☐ 5 COARSE | ☐ 12 PROCESSED | ☐ 2 REG GLASSES | ☐ 12 FRONT TORSO | _____ | ☐ 12 BODILY FORCE _____ |
| ☐ 6 WIRY | ☐ 13 TEASED | ☐ 3 SUN GLASSES | ☐ 13 BACK TORSO | _____ | ☐ 13 STRANGULATION _____ |
| ☐ 7 WIG | ☐ 14 OTHER_____ | ☐ 4 WIRE FRAME | ☐ 14 OTHER | _____ | ☐ 14 TIRE IRON _____ |
| ☐ 8 OTHER_____ | **FACIAL HAIR 9** | ☐ 5 PLASTIC FRAME | | _____ | ☐ 15 OTHER _____ |
| **HAIR CONDITION 6** | ☐ 0 UNKNOWN | ☐ Color_____ | | | **WEAPON FEATURE 18** |
| ☐ 0 UNKNOWN | ☐ 0 N/A | ☐ 6 OTHER_____ | **UNIQUE CLTHNG 15** | **WEAPON IN 16** | ☐ 0 UNKNOWN  ☐ 0 NONE |
| ☐ 1 CLEAN | ☐ 1 CLN SHAVEN | **VOICE 12** | ☐ 0 UNK  ☐ 0 NONE | ☐ 0 UNKNOWN | ☐ 1 ALTERED STOCK _____ |
| ☐ 2 DIRTY | ☐ 2 MOUSTACHE | ☐ 0 UNKNOWN | ☐ 1 CAP/HAT | ☐ 0 N/A | ☐ 2 SAWED OFF _____ |
| ☐ 3 GREASY | ☐ 3 FULL BEARD | ☐ 0 N/A | | ☐ 1 BAG/BRIEFCASE | ☐ 3 AUTOMATIC _____ |
| ☐ 4 MATTED | ☐ 4 GOATEE | ☐ 1 LISP | ☐ 2 GLOVES | ☐ 2 NEWSPAPER | ☐ 4 BOLT ACTION _____ |
| ☐ 5 ODOR | ☐ 5 FUMANCHU | ☐ 2 SLURRED | | ☐ 3 POCKET | ☐ 5 PUMP _____ |
| ☐ 6 OTHER_____ | ☐ 6 LOWER LIP | ☐ 3 STUTTER | ☐ 3 SKI MASK | ☐ 4 SHOULDER | ☐ 6 REVOLVER _____ |
| **R/L HANDED 7** | ☐ 7 SIDE BURNS | ☐ 4 ACCENT | _____ | HOLSTER | ☐ 7 BLUE STEEL _____ |
| ☐ 0 UNKNOWN | ☐ 8 FUZZ | Describe_____ | ☐ 4 STOCKING MASK | ☐ 5 WAISTBAND | ☐ 8 CHROME/NICKEL _____ |
| ☐ 1 RIGHT | ☐ 9 UNSHAVEN | _____ | _____ | ☐ 6 OTHER_____ | ☐ 9 DOUBLE BARREL _____ |
| ☐ 2 LEFT | ☐ 10 OTHER_____ | ☐ 5 OTHER_____ | ☐ 5 OTHER_____ | | ☐ 10 SINGLE BARREL _____ |
| | | | | | ☐ 11 OTHER _____ |

_____
_____
_____
_____
_____
_____
_____
_____
_____

ESPD Form #200 (Rev 5/97)

**Figure 4-2** (Courtesy of the El Segundo Police Department)

Just as in the case of a felony, when a misdemeanor crime report is being completed, a person should be named as a suspect only when sufficient information exists to justify their arrest. In most jurisdictions a peace officer cannot arrest a person for a misdemeanor committed outside his or her presence. Because of this, it would generally not be a problem to list someone as a misdemeanor suspect even if this level of probable cause was absent; but, the successful investigative report writer is striving to develop consistent habits and ways of doing things. If this consistency is developed the task of writing reports is made much easier. One way in which this report writing simplification will be readily apparent is to have a common method for listing suspect information regardless of the circumstances.

### Witnesses

The test to determine who a witness might be is an entirely different matter. A witness is someone who has useful information about a crime. Ideally, a witness would have this useful information because they became aware of it through one of the five senses: seeing, hearing, touching, tasting, or smelling something. This, however, is not always the case. A witness may be someone who is aware of evidence of a crime because they saw it occur or because someone who witnessed the crime told them where evidence might be located. When a person is found who has useful information about a crime being investigated they should be listed as a witness. Useful information may be defined as anything that helps solve a crime, points to or eliminates a particular person as being involved, identifies property or evidence in a given case, or anything else that might be important to the investigation.

### Others

Not all people an investigator talks to during an investigation can be categorized in one of the areas already discussed. There will be many people who have no information, who did not see or hear anything at all relative to the matter being investigated, and whose only reason for being contacted by the investigator is that they were in the same area as the investigator at a time he or she wanted to talk to them. Persons in this category are known as OTHERS in report writing vernacular. Not only is there a good reason for including them in the report, but there is also a proper way to do so. An example involving an OTHER is:

> A car parked at a shopping mall is vandalized by someone who throws a brick onto the hood, denting the metal and scratching the paint. The police are called and while the officer talks to Robert Smith, the owner of the victimized car, Brenda Wilson, who owns the car parked next to Smith's, arrives at her car. The officer asks Wilson if she saw anything unusual when she arrived at the mall. Wilson tells the officer that she arrived two

hours before Smith did and was shopping inside the mall when Smith arrived.

The question is, what does the officer do with the information from Wilson? There are officers who would not include Wilson's information in the report and there are others who would make Wilson a witness; but, the primary question is this, *Does Wilson have any useful information about the case?* The answer is clearly no, she does not. The next logical question is, *How should Wilson's information be handled in the report?* The correct method of handling it is to include it in the narrative in the chronological order in which it was received. It might read:

> While I was talking to Smith, Brenda Wilson, 123 Maine, Pasadena, CA, 90876, (818) 555-1212, arrived and told me . . . .

Why is it important to include this information in the report? For the following reasons:

1. It is appropriate to document all of the investigation you do. Your supervisors might review your reports to see what type of investigations you do, as well as evaluate their quality, thoroughness, and the manner in which you proceed.
2. Even if you merely talk to someone you should put it in the report. Even though the person tells you they did not see or hear anything related to the inquiry, it may turn out that they were somehow involved.
3. It is a way to document what you did and this adds to your credibility if you are called to testify in court.
4. The person you are talking to who claims to know nothing about the crime may be lying about their involvement, although you may be unaware of it at the time. Once the report is forwarded to the investigation division for follow up, the name of the uninvolved person may mean something to the other investigators who read or hear about the case.

It is important to *keep the big picture in mind* when you are investigating what seems like a small matter, because more than once a person who seems uninvolved today is a suspect tomorrow.

Describing people in a report is a simple matter if the witness can provide a good description. If, however, no one saw the suspect, or saw him or her for a split second, getting a good description may be more difficult and presents a challenge to you, the investigator.

## Assessing Weights and Measurements

To start, you must have a basic ability to assess weights and measurements. For example, you should know how tall you are, as well as the distance from the ground to the tip of your nose, to the top of your shoulder, and to your waist. You should know your weight and the approximate weights of people at various heights with average builds. With this knowledge you can practice guessing the height and weight of people you know until you become accurate. Perfecting this skill will serve you well as you interview witnesses and ask them about people they saw committing crimes you are investigating.

## Interviewing for Suspect Descriptions

As you interview people to get suspect descriptions, it may not be realistic to expect a complete description on the first try. What is more likely to occur is that they will tell you the things they remember. The information they will remember will include the things that stood out about the person in the order of importance to the witness. The order in which they describe the suspect to you will make perfect sense to them, but it will probably not be in the order you need to complete your report. What good investigators will do is have a plan in place that allows them to get the information they need in a timely fashion and allows the witness to report all the information he or she knows.

Now is a good time to remember the basic rule of note taking which is **listen first, then write.** Ask the witness to tell you what the suspect looked like and as he or she tells you, just listen. Once finished, ask him or her to answer specific questions based on the order of information you need for your report. A suggested order or formula for recording this information is:

Sex, race, age, height, weight, hair style and color, eyes, clothing description starting at the top with outside garments and working down, and anything else that is of importance such as tattoos, missing limbs or teeth, accents, unusual gait, etc.

People will not always recall all the information about a suspect even when they had an unobstructed view of the person for several minutes. In those instances where you are able to get a full and complete description it is a simple matter to write it into the report. In those cases where several parts of the description are left unfilled it is acceptable to continue the description with the next piece of known information. For example:

Male, white, about 35, 6'3", 197 pounds, short brown hair, wearing a white short sleeve sweatshirt, gray shorts, white slip-on shoes.

As you can see, there is a great deal of information missing from this description as compared to the ideal formula, yet there is enough information available to begin a search. A description is used not only to help find the suspect, but also to eliminate those who are not involved from the suspect pool.

There will be times when a witness provides information that is too lengthy to fit into the boxes on the crime report face sheet. In these cases write SEE NARRATIVE in the suspect information boxes and then write the suspect description on the first page of the narrative after labeling the area SUSPECT INFORMATION. This will give you all the room you need to write a thorough suspect description. Keep in mind that the spaces on the face sheet are designed to guide and help you complete the report, not limit what you write. When it comes to describing suspects, you can never have too much good information.

In the event you have multiple suspects start by helping the witness focus on one suspect and get as complete a description as possible before repeating the same procedure for all additional suspects. If you have no names for the suspects, it is acceptable and appropriate to give them numbers such as suspect #1 and suspect #2.

It is rare that several witnesses see a suspect and describe him or her in the same terms. In those cases where witnesses give different descriptions you should list the suspect by witness description so that it is clear which witness said what about the suspect. For example:

Suspect #1 as described by Cleworth
    male, white, 35 years, 6'5", 230 pounds, black hair . . .
Suspect #1 as described by Najjar
    male, hispanic, 30 years, 6'3", 215 pounds, dark hair . . .
Suspect #1 as described by Ito
    male, white, mid 30s, 6'4", 220 pounds, black hair . . .

In this case the witnesses saw a person they believe to be the suspect but all remember him differently. It would be unacceptable to combine the information into one hybrid description for the sake of speed and brevity. This would not help solve the crime and quite possibly would hinder any prosecution if the witnesses became confused about what they saw.

## DESCRIBING PROPERTY IN A REPORT

How to record thorough and accurate property descriptions poses two main problems for the investigative report writer. First, the many different brands, sizes, colors, and models of products available create an almost infinite number of things to describe. Second, recording an accurate value for this infinite number of items is difficult because the cost can be influenced by appreciation, depreciation, damage, or collectablilty.

# LA HABRA POLICE DEPARTMENT    PROPERTY REPORT    Page _____ of _____

150 N. Euclid, La Habra, CA 90631
(562) 905-9750   Fax (562) 905-9779

[ ] ORIGINAL REPORT    [ ] SUPPLEMENTAL REPORT    CASE NO.

| ITEM No. | TYPE | QUANTITY | ARTICLE | BRAND | MODEL | SERIAL NUMBER | VALUE |
|---|---|---|---|---|---|---|---|
| | COLOR (S) | | | GENERAL DESCRIPTION | | | |
| | PREMISES / AREA / ROOM TAKEN FROM | | | WHERE IS PROPERTY NOW? | | ADDITIONAL NOTES | |

| ITEM No. | TYPE | QUANTITY | ARTICLE | BRAND | MODEL | SERIAL NUMBER | VALUE |
|---|---|---|---|---|---|---|---|
| | COLOR (S) | | | GENERAL DESCRIPTION | | | |
| | PREMISES / AREA / ROOM TAKEN FROM | | | WHERE IS PROPERTY NOW? | | ADDITIONAL NOTES | |

| ITEM No. | TYPE | QUANTITY | ARTICLE | BRAND | MODEL | SERIAL NUMBER | VALUE |
|---|---|---|---|---|---|---|---|
| | COLOR (S) | | | GENERAL DESCRIPTION | | | |
| | PREMISES / AREA / ROOM TAKEN FROM | | | WHERE IS PROPERTY NOW? | | ADDITIONAL NOTES | |

| ITEM No. | TYPE | QUANTITY | ARTICLE | BRAND | MODEL | SERIAL NUMBER | VALUE |
|---|---|---|---|---|---|---|---|
| | COLOR (S) | | | GENERAL DESCRIPTION | | | |
| | PREMISES / AREA / ROOM TAKEN FROM | | | WHERE IS PROPERTY NOW? | | ADDITIONAL NOTES | |

| ITEM No. | TYPE | QUANTITY | ARTICLE | BRAND | MODEL | SERIAL NUMBER | VALUE |
|---|---|---|---|---|---|---|---|
| | COLOR (S) | | | GENERAL DESCRIPTION | | | |
| | PREMISES / AREA / ROOM TAKEN FROM | | | WHERE IS PROPERTY NOW? | | ADDITIONAL NOTES | |

| ITEM No. | TYPE | QUANTITY | ARTICLE | BRAND | MODEL | SERIAL NUMBER | VALUE |
|---|---|---|---|---|---|---|---|
| | COLOR (S) | | | GENERAL DESCRIPTION | | | |
| | PREMISES / AREA / ROOM TAKEN FROM | | | WHERE IS PROPERTY NOW? | | ADDITIONAL NOTES | |

| ITEM No. | TYPE | QUANTITY | ARTICLE | BRAND | MODEL | SERIAL NUMBER | VALUE |
|---|---|---|---|---|---|---|---|
| | COLOR (S) | | | GENERAL DESCRIPTION | | | |
| | PREMISES / AREA / ROOM TAKEN FROM | | | WHERE IS PROPERTY NOW? | | ADDITIONAL NOTES | |

**PROPERTY TYPES:    S = STOLEN ;  R = RECOVERED ;  SR = STOLEN AND RECOVERED ;  SK = SAFE KEEPING ;  F = FOUND ;  E = EVIDENCE**

| REPORTING OFFICER / ID # | | DATE | APPROVED BY | | DATE |
|---|---|---|---|---|---|
| COPIES TO: | [ ] CHIEF  [ ] NFP  [ ] PATROL  [ ] DET  [ ] TRAFFIC<br>[ ] DMV  [ ] CAU  [ ] ABC (2)  [ ] DA  [ ] OTHER _____ | | ROUTED BY | ENTERED BY | |

**Figure 4-3**   (Courtesy of the La Habra Police Department)

## The Average Person Test

Describing property is probably the easier of the two tasks addressed in this chapter. There is a simple way to get a quality description of any piece of property, whether it is something that has been reported as stolen by a victim, booked as a piece of evidence, turned in as found property, or suspected of being contraband but its status cannot be determined on the spot.

This simple method is known as the **Average Person Test** and is easy for the investigator to apply. Whenever you are describing any piece of property or evidence, do so with the objective of writing a description that is so complete and so thorough that the average person could look at five or six similar items and, based on the description you wrote, pick out the item you have described. If the average person can do this then the description is good enough. If the average person cannot, the description is not good enough.

## Determining Property Value

Assigning a value to stolen property is another area of concern for many investigators because there are several ways to do so. You can spend hours haggling with victims about the value of their five-year-old compact disc player, or the intrinsic value of the coin collection they have had since they were a small child, but it is necessary to have a value for property because the value of the loss is important in determining the corpus, as in the case of misdemeanor or grand theft. The amount of the loss in theft cases is also used in determining the level of assignment for investigators. Limited resources may prevent some agencies from working theft cases with a minimal loss as diligently as they would a case with a large one.

Fortunately, there are many ways of arriving at value estimates in investigative reports. Some of the most common methods are:

1.  *Original cost.* Using this method requires the victim or owner of the property to remember what they paid for the item. This amount is used in the report without regard for subsequent damage that might have lessened the value, or to appreciation that might have increased the value. This method sometimes requires the owner to present receipts showing the purchase price.
2.  *Fair market value.* This method requires the investigator to use his or her judgment in determining the value of an item by considering what it was worth when it was new, what the current demand for the item is, and what it may be worth now. Interest by the public or of some collectors group might increase the value. The rarity of the item might also be a factor. When using this method the investigator makes a judgment based on his or her best-informed guess as to the value and includes this in the report.

## EL SEGUNDO POLICE DEPARTMENT
# PROPERTY REPORT

DATE AND TIME OF REPORT

DR#

- ☐ ARRESTEE
- ☐ SUSPECT
- ☐ VICTIM
- ☐ WITNESS

- ☐ EVIDENCE
- ☐ FOUND PROPERTY
- ☐ SAFEKEEPING
- ☐ PRISONER PROPERTY

- ☐ DESTRUCTION
- ☐ UNDER OBSERVATION
- ☐ HOLD: CONTACT OFFICER _____
- ☐ OTHER _____

NAME (Property Booked To)

NAME (Secondary Persons)

☐ MISDEMEANOR
☐ FELONY

ADDRESS

ADDRESS

**CHARGES**

1.

CITY                    ZIP

CITY                    ZIP

2.

PHONE
RES.:              BUS.:

PHONE
RES.:              BUS.

3.

**BIKE**

| MAKE | MODEL | SPEED ☐ BOY'S  ☐ GIRL'S  1 ☐  3 ☐  5 ☐  10 ☐  15 ☐  OTHER ☐ | FRAME COLOR | WHEEL SIZE |

SERIAL NO.        LICENSE NO.      CITY       OTHER DESCRIPTIONS

**FOUND PROPERTY STATEMENT:** DO YOU WISH TO CLAIM THIS PROPERTY IF THE LEGAL OWNER CANNOT BE LOCATED?   ☐ YES   ☐ NO
PROPERTY CAN BE CLAIMED AFTER **90 DAYS** DATE.................   FINDER SIGNATURE...................................................

**PRISONER PROPERTY STATEMENT:** ITEMS NOT ABLE TO GO TO COURT WILL BE HELD BY THIS DEPARTMENT FOR A PERIOD OF **ONE YEAR.** IF AFTER 1 YEAR YOU HAVE NOT MADE ARRANGEMENTS WITH THE PROPERTY & EVIDENCE OFFICER TO CLAIM YOUR PROPERTY WE WILL DESTROY ALL PRISONER PROPERTY AFTER THAT DATE.
**PRISONER SIGNATURE** ................................... DATE.................

LIST ALL ARTICLES STARTING WITH ITEM NO. 1: DESCRIBE PROPERTY AS FOLLOWS.

| ITEM NO. | QTY. | ARTICLE, DESCRIPTION, BRAND NAME, MODEL, ETC. | SERIAL NO. | VALUE OR WEIGHT |
|----------|------|-----------------------------------------------|------------|-----------------|
|          |      |                                               |            |                 |
|          |      |                                               |            |                 |
|          |      |                                               |            |                 |
|          |      |                                               |            |                 |
|          |      |                                               |            |                 |
|          |      |                                               |            |                 |
|          |      |                                               |            |                 |
|          |      |                                               |            |                 |
|          |      |                                               |            |                 |

CONTINUATION FORM USED FOR ADDITIONAL ITEMS ☐

**COMPLAINT FORM NARRATION:**

REPORTING OFFICER (Name and Serial No.)

TEMPORARY LOCATION OF PROPERTY

APPROVING SUPERVISOR        DATE

PROPERTY USE ONLY
PROPERTY ROOM LOCATION

BY:

DATE

FINAL DISPOSITION

PROPERTY RELEASED TO:

ADDRESS:

DATE

**WHITE:** WITH EVIDENCE        **YELLOW:** TO RECORDS        **PINK:** TO FINDER/PRISONER PROPERTY

**Figure 4-4**   (Courtesy of the El Segundo Police Department)

## CONTINUATION REPORT

| DATE | | OFFICER | | | DR# | |
|---|---|---|---|---|---|---|

| ITEM NO. | QTY. | ARTICLE DESCRIPTION, BRAND NAME, MODEL ETC. | SERIAL NO. | VALUE |
|---|---|---|---|---|
| | | | | |
| | | | | |
| | | | | |
| | | | | |
| | | | | |
| | | | | |
| | | | | |
| | | | | |
| | | | | |
| | | | | |
| | | | | |
| | | | | |
| | | | | |
| | | | | |
| | | | | |
| | | | | |
| | | | | |
| | | | | |
| | | | | |
| | | | | |
| | | | | |
| | | | | |
| | | | | |
| | | | | |
| | | | | |
| | | | | |
| | | | | |
| | | | | |
| | | | | |
| | | | | |

| REPORTING OFFICER (Name and Serial No.) | TEMPORARY LOCATION OF PROPERTY | APPROVING SUPERVISOR | DATE |
|---|---|---|---|
| PROPERTY USE ONLY PROPERTY ROOM LOCATION | BY: | DATE | FINAL DISPOSITION |
| FCN NO. | BY: | | DATE |

**Figure 4-5**   (Courtesy of the El Segundo Police Department)

3.   *Victim appraisal.* This method requires the victim to provide the value of the missing property by telling the investigator how much it is worth. The victim may use, among other criteria, what they perceive as the replacement cost, historical value, or estimated replacement cost, but the bottom line is that the value is going to be what the victim says. This is a less than scientific way of determining the value of something, but investigators waste a lot of valuable time determining the value of property when it really is not that important to the investigating agency. Unless the value is over-reported by thousands of dollars the net effect on the investigating agency is minimal. The agency is still going to investigate the crime and perform the administrative duties required by law. If the real concern is to report an accurate amount to prevent an insurance fraud, it is necessary to consider that insurance companies utilize the services of agents, private investigators, and estimators to verify the value of property taken before paying a claim. These claims adjusters, not the investigator, have as one of their primary responsibilities the duty of determining the value of a loss.

4.   *Replacement cost.* Figuring a replacement cost can be tricky. Not only does the investigator need to know where an item can be purchased, but also if there is any kind of rebate plan, sale price, or discount available to the victim.

There are advantages and disadvantages to each of these methods of property value determination. In choosing the method that will be used it is important to consider agency requirements. If a particular method is used by the agency, the investigator should stay with that method for the sake of consistency. If the investigator has the option of selecting a method, the victim appraisal method is recommended because it quickly establishes a value. If the victim over-estimates the value of the loss, the insurance company will correct them at the time the settlement is made and your report can be adjusted accordingly.

Determining the value of property should not be the most involved or important thing an investigator does during a preliminary investigation. It should take a minimal amount of time and allow him or her to move forward in trying to locate witnesses and physical evidence.

## WRITING EVIDENCE REPORTS

Describing property in crime reports is only one application in which you must be accurate and specific in your writing. Another report writing area where excellence in writing is the only acceptable standard is when dealing with evidence in the aptly titled EVIDENCE REPORT. An evidence report has two basic components. The first is a **description of the evidence** and the second is the **chain of custody.**

Evidence can take many forms and come into your care and custody in a variety of ways. If something tends to prove a fact or gives you a basis for believing something, it is likely evidence. How you identify and collect evidence in your specific cases is the subject of another class and another skill set. For the purpose of this book, let's assume that your skills are good and you are using the proper techniques. Now your problem is introducing the evidence into your report and accounting for it as the case continues.

There are many ways to account for evidence but the two most popular are to refer to the evidence in the narrative of the investigative report in the chronological order it was received, using great detail and accountability; and the other is to refer to the evidence in the narrative in a generic sense and then in great detail in a separate report called the Evidence Report. The second method is recommended because it gives greater freedom to the writer in describing the evidence, greater organization in the report, and it makes the report easier to read and understand.

## Describing Evidence

Just as with property descriptions, there is no room for error when describing evidence. The description you use in the evidence report must also meet and exceed the **average person test.** The level of additional detail is needed because you may be asked to identify a specific piece of evidence from the witness stand while someone holds or displays the item several feet from you. This might seem like a big problem, but it is easy to solve if proper marking and packaging techniques are used.

The marking system you select and use should not destroy the evidence or diminish its value. It just needs to be something you can identify as your mark at a later date. The packaging and tagging process should be something that is manageable with as little effort as possible and still give a solid level of protection to the evidence. Packaging materials are commonly found in the evidence preparation rooms of police stations.

## Evidence Report Formats

While there are many different crime report face sheets, there are basically two different report formats used for evidence reports. The first format uses a printed form that serves much like a crime report face sheet. It organizes the information and prompts the writer with headings. This is in essence a fill-in-the-blanks report. Although there are differences in the way the form looks from one agency to another, the information needed to complete it is basically the same.

The second format uses a free-flowing narrative on a blank piece of paper, which allows the writer to include as much detail and information about the evidence as is needed. Some agencies have strict guidelines about which items of evidence to list first, second, third, and so on. There are a number of agencies that do not have a protocol for evidence reports. In these cases a workable solution to

# El Fuego Police Department

**Evidence Report**                              **99-076587**

    On 4-27-99 at 1545 hours a search warrant was served at
623 Fastwater Lane.  I found the following items during the
subsequent search.

1.    (1) Walther PPK, .380 Caliber handgun, blue steel finish
      with brown plastic grips.  Serial number WP164128.
      Found under couch in the living room.

2.    $647.00 U.S. currency consisting of (5) $100 bills, (2)
      $50 bills, (4) 10 bills, and (7) $1.00 bills.  Found
      under bed in master bedroom.

3.    $1000 U.S. currency consisting of (10) $100 bills.
      Found in freezer wrapped in foil.

4.    (1) clear plastic baggy containing an unknown type
      white powder.  463 grams gross weight.

5.    (1) clear plastic baggy containing marijuana.  1000
      grams gross weight.

I kept all the items with me and marked, packaged, tagged,
and booked them at the El Fuego Police Department.

    M. Colin     #167

**Figure 4-6**

the problem is to divide the evidence into general categories and
then list them in the following order:

> GUNS
>
> MONEY
>
> DRUGS
>
> ITEMS WITH SERIAL NUMBERS
>
> ITEMS WITHOUT SERIAL NUMBERS

These headings would not appear on the evidence report but
rather, act as a guide for you in setting up the report. The items on
the evidence report should be consecutively numbered beginning

| PROPERTY OF: | | |
|---|---|---|
| DATE: | | DR# |
| DENOM. | NUMBER | AMOUNT |
| 100's | | $ |
| 50's | | |
| 20's | | |
| 10's | | |
| 5's | | |
| 1's | | |
| OTHER | | |
| CHANGE | | |
| TOTAL | | $ |
| OFFICER'S SIGNATURE | | |
| VERIFYING OFFICER | | |
| SUBJECT'S SIGNATURE | | |

PE-028

**Figure 4-7**  (Courtey of the El Segundo Police Department)

with the first item seized. This system will allow additional evidence to be added if the case is ongoing. The particulars of where and when the evidence was seized would be shown in each item.

No matter which format of evidence report you use, you must account for all the evidence you seize, in other words, you must establish a **chain of custody.**

## Establishing the Chain of Custody

The chain of custody is the term that describes the handling and care of evidence. When evidence is seized, it must be identified, described, and accounted for from the time it was seized until it is entered into evidence at trial.

Typically, the process or path that evidence takes begins with the investigator finding an item that has real or potential evidentiary value. The item is collected, processed if necessary, and then marked for later identification. Once marked, the item is packaged, tagged, and placed in a secure evidence storage facility. Once it has been

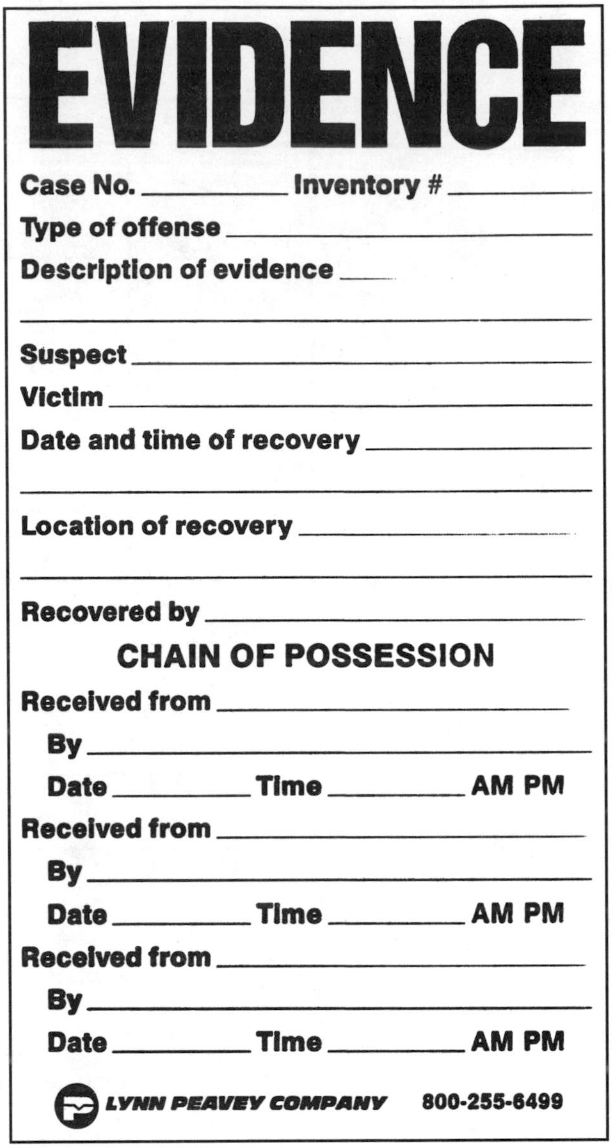

# EVIDENCE

Case No. ________ Inventory # ________

Type of offense ________

Description of evidence ________

________

Suspect ________

Victim ________

Date and time of recovery ________

________

Location of recovery ________

________

Recovered by ________

**CHAIN OF POSSESSION**

Received from ________

By ________

Date ________ Time ________ AM PM

Received from ________

By ________

Date ________ Time ________ AM PM

Received from ________

By ________

Date ________ Time ________ AM PM

LYNN PEAVEY COMPANY 800-255-6499

**Figure 4-8** (Courtesy of the El Segundo Police Department)

booked in the evidence facility, the responsibility to care for the evidence rests with the person who manages the facility or laboratory where you put the evidence.

Establishing the chain of custody is also part of your responsibility as the report writer, and it is properly done in the evidence report. The report numbers each piece of evidence separately beginning at one, and includes a description of the item, where it was found, who found it, and where it was booked. In general, this establishes the chain of custody and might look like:

> On 10-13-98 at 1130 hours, I found the following item at 640 Peach Street and marked, packaged, tagged, and booked it at the Department of Justice Lab in Sacramento.
>
> 1. One gallon red plastic gasoline can, empty, found in the garage.

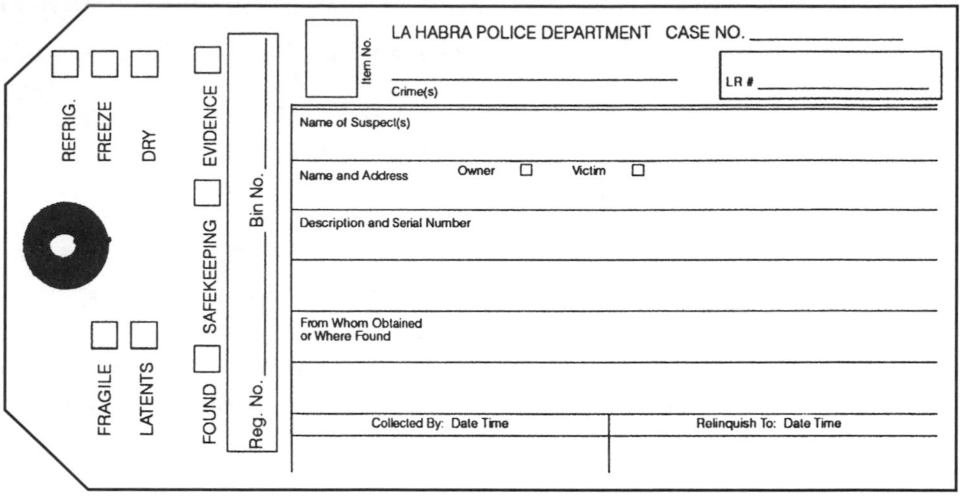

**Figure 4-9**   (Courtesy of the La Habra Police Department)

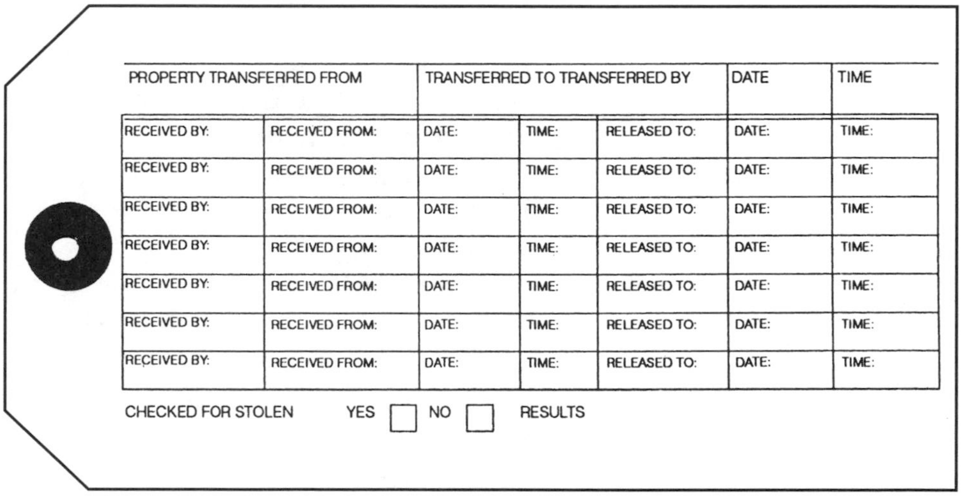

**Figure 4-10**   (Courtesy of the La Habra Police Department)

This example can be expanded to any number of items found at the same scene and allows the writer to keep the chain of custody clear and well established. This also shows the benefit of writing in the active voice.

You may be involved in cases where several investigators search for and find evidence and then give it to another investigator who is responsible for the report. In this kind of case the items would be numbered and described just as in the first example, with the addition of who found it. The *chain of custody* would look like this:

On 12-25-99 at 1845 hours a search was conducted at 123 Main Street and the following evidence was found:

1. (1) one gallon red plastic gasoline can, empty, found in the garage by Ho.
2. (1) partially burned book of matches found in the kitchen by Fava.

3.   (1) 3′ by 4′ piece of carpet, stained with an unknown liquid, found in the hallway by Conklin.

I collected each piece of evidence from the finder, marked it, packaged it, tagged it, and booked it at the El Fuego Police Department evidence room.

This example will accommodate any number of searchers and any number of evidence items. It can expand or contract as needed to establish good control of the evidence. The key to success with any type of evidence is to properly handle and care for it, and be able to account for it from the moment you seize it until you testify about it in court.

## REVIEW

1.   The people in a report are defined by what role they have. Only list someone as a felony suspect when there is probable cause to arrest them.
2.   A witness is someone with useful information.
3.   The name of everyone you talk to should be included in the report.
4.   Property descriptions must pass the "average person test."
5.   Four methods of determining the value of stolen property are:
Original cost method
Fair market value method
Victim appraisal method
Replacement cost method
6.   Evidence reports can be free-flowing or fill in the blanks. A format for the free-flowing style is:

GUNS

MONEY

DRUGS

ITEMS WITH SERIAL NUMBERS

ITEMS WITHOUT SERIAL NUMBERS

7.   The chain of custody must be established.

## EXERCISES

1.   Measure and commit to memory the height of three reference points on your body.
2.   Using these reference points, estimate the height and weight of five persons in your class. Were you close? Discuss why and how you can improve in this area.

Chapter 4  Describing Persons and Property    59

3.  Assume the classroom you are in was burglarized the past weekend and a color television set belonging to the school was taken along with a VCR belonging to the class instructor. Who are the victims? Explain your answer.

4.  Write a description of the class instructor using the formula for describing persons.

5.  Locate three articles in the classroom and describe each of them so that the descriptions pass the average person test.

6.  Design a property report form.

7.  Using the property report form, make an inventory of the appliances and electronic and entertainment equipment in your home.

8.  Using the items in Exercise 5, prepare an evidence report that includes a chain of custody. You are the finder of all three items and you booked them at the Big Pine Police Department.

9.  Using the same scenario as Exercise 8, change the chain of custody to show three of your classmates as the finders and you as the person who collected and booked the items.

## QUIZ

1.  What is the definition of a victim?
2.  What is the definition of a witness?
3.  When should a person be named as a suspect in a felony crime report?
4.  What is the order or formula for describing people in a report?
5.  How should clothing be described?
6.  If you have multiple witnesses and each provides a suspect description, when should the descriptions be combined into a composite description?
7.  If a witness provides so much suspect information that it will not fit on the face page, where should it go?
8.  When describing stolen property in a report, how good should the description be?
    a.  Completely accurate
    b.  Only list the make, model, and serial number
    c.  So it can pass the average person test
    d.  As detailed as possible in 30 words or less
9.  Describe the original cost method of determining property value.
10. What is the chain of custody?

# 5 Crime Reports

## K E Y  P O I N T S

What is a crime report

The purpose of a crime report

The components of a crime report

What must be established in a crime report

The purpose of a crime report face sheet

Commonly needed information to complete a crime report

The importance of the preliminary investigation cannot be overly emphasized because it forms the foundation for further investigative efforts and, for many crimes, represents the only investigation that is done. As such, it is imperative that the investigation be as thorough as possible and for the crime report to be clear, concise, accurate, and complete. The successful outcome of a case largely depends upon the quality of the information collected during the preliminary investigation.

## Purpose of a Crime Report

Of the hundreds of different reports used by the investigative organizations in this country, the most common is the crime report. A crime report can have many purposes depending on who is using it, but its application in the investigative process is well-defined. First and foremost, the purpose of a crime report is to document that a crime has been reported. The most important thing to be established in a crime report is the corpus delicti. If the corpus is not established there is no need for a crime report. This is not to say that if there is no crime then there is no need for a report of some kind, but rather, that the crime report is not the best way to document the incident. There is normally a great need to document the events an agency investigates, and in most departments there is an appropriate method to do so other than with a crime report. Although the requirements for completing a crime report vary from law enforcement agency to law enforcement agency, it is a good rule of thumb to say that if the elements of a crime are present, a crime report should be completed. Some law enforcement personnel may challenge this point of view by saying that a crime report is not necessary if the victim is not desirous of prosecution; however, valuable information and evidence may be lost or become irretrievable if not documented or collected at the time.

Second, even though the victim may not be interested in moving forward with a prosecution at that point, he or she may change his or her mind. Third, the decision to prosecute someone does not rest with the officer or investigator, but rather with the district attorney's office. Last but not least, documenting a crime when one is indicated allows an investigative agency to see the big picture with regard to activity and crime in the area. This knowledge allows the agency to plan deployment and resource strategies accordingly. This last point bears further note because almost without exception law enforcement managers forecast deployment plans and staffing levels based in great part on the number of documented crimes occurring within certain time frames in a geographic area.

Completing the report as soon as possible eliminates the need for another officer to return later and do so. This wasteful practice of sending an investigator out to the scene a second time not only takes time and costs money, but lowers morale and fosters ill feelings among the officers having to do the work previously assigned to others.

Law enforcement agencies use crime reports for many purposes including the identification of suspects, listing stolen property, establishing methods of operation being used by area criminals, determining when crimes occurred in order to staff appropriately, documenting statistics for a multitude of reasons, and as a way of justifying the arrest of those believed to have committed the crimes. The most important use of a crime report, however, is as an **investigative tool.**

Included in the responsibility law enforcement has to the community it serves is the role of helping those in need. Another major part of this responsibility is finding and bringing criminals to justice. It is difficult, if not impossible, to investigate these incidents of criminal wrong-doing without adequate crime reporting and an initial investigation.

Investigators who are charged with the task of completing crime reports must recognize that when they are assigned a radio call of a burglary or petty theft they are not being assigned to go to the location and take a report; rather, they are being given the opportunity to investigate the incident and report what they learn. As discussed in an earlier chapter, an investigator is bound only by his or her imagination, energy level, and the law.

For all the similarity in purpose and goals among the investigative agencies in the United States, there are differences in the manner and style of crime reporting. Just as each agency has its own uniform for its investigators and color combinations for its patrol cars, each has its own method and configuration of crime report documentation.

## Completing Crime Reports

Although there is a difference in the way these agencies complete crime reports, and in the configuration of their crime report face sheets, the face sheets and accompanying free-flowing narrative sections are relatively simple in design and use. There are two basic parts to a crime report: the **face sheet** and the **narrative section** (as described in chapter 3), which contains the investigation. To simplify the task of completing the face sheet it is helpful to understand its purpose. Despite the wide variety of configurations and boxes present on the thousands of face sheets in use by this country's investigative agencies, the purpose of a crime report face sheet is twofold. One is to *organize information* and the other is *to gather statistics.*

Once the investigator is familiar with the information that is requested on the face sheet, the style of interviewing can be adjusted so that the questions asked of the victim or witness are in the same order they occur on the face sheet. This will reduce the time needed to *fill in the blanks.* In spite of the different crime

| | | | | | A ☐ ACTIVE | CASE NUMBER | |
|---|---|---|---|---|---|---|---|
| ☐ NO PROSECUTION DESIRED | **EL SEGUNDO POLICE DEPARTMENT** | | | | S ☐ SUSPENDED | | |
| ☐ TELEPHONE REPORT | 348 MAIN STREET | | | | R ☐ RECORDS | | |
| ☐ INSURANCE REPORT | EL SEGUNDO, CA 90245 | | | | C ☐ CLOSED | REFER OTHER RPTS | |
| ☐ COURTESY REPORT | 310-322-9114 | | | | K ☐ COURTESY | | |
| ☐ DOMESTIC VIOLENCE | **CRIME REPORT** | | | | U ☐ UNFOUNDED | | |
| ☐ CONFEDENTIAL SEX CRIME | | | | | | | |

**CRIME**

| CODE SECTION | CRIME | | | | UCR CODE | SECONDARY-COUNTS | OTHER-COUNTS |
|---|---|---|---|---|---|---|---|
| SPECIFIC LOCATION OF CRIME | | | | OCCURRED ON/OR BETWEEN: | DATE | DAY | TIME |
| BUSINESS NAME | | DATE RPT'D | TIME RPT'D | AND: | DATE | DAY | TIME |

**VICTIM**

| NAME (Last, First, Middle) | | OCCUPATION | | D.O.B. | AGE | SEX ☐ 1. M ☐ 2. F | RACE ☐ 1. WHT ☐ 2. HISP ☐ 3. BLK ☐ 5. CHI ☐ 7. FIL ☐ 9. P.ISL. ☐ 4. IND ☐ 6. JAP ☐ 8. OTH.____ |
|---|---|---|---|---|---|---|---|
| RESIDENCE ADDRESS | | | CITY | ZIP CODE | | RES. PHONE ( ) | |
| BUSINESS NAME AND ADDRESS | | | CITY | ZIP CODE | | BUS. PHONE ( ) | |

**VICTIM(S) - WITNESS - RP**

| CODE | NAME (Last, First, Middle) | | OCCUPATION | | D.O.B. | AGE | SEX ☐ 1. M ☐ 2. F | RACE ☐ 1. WHT ☐ 2. HISP ☐ 3. BLK ☐ 5. CHI ☐ 7. FIL ☐ 9. P.ISL. ☐ 4. IND ☐ 6. JAP ☐ 8. OTH.____ |
|---|---|---|---|---|---|---|---|---|
| RESIDENCE ADDRESS | | | | CITY | ZIP CODE | | RES. PHONE ( ) | |
| BUSINESS NAME AND ADDRESS | | | | CITY | ZIP CODE | | BUS. PHONE ( ) | |
| CODE | NAME (Last, First, Middle) | | OCCUPATION | | D.O.B. | AGE | SEX ☐ 1. M ☐ 2. F | RACE ☐ 1. WHT ☐ 2. HISP ☐ 3. BLK ☐ 5. CHI ☐ 7. FIL ☐ 9. P.ISL. ☐ 4. IND ☐ 6. JAP ☐ 8. OTH.____ |
| RESIDENCE ADDRESS | | | | CITY | ZIP CODE | | RES. PHONE ( ) | |
| BUSINESS NAME AND ADDRESS | | | | CITY | ZIP CODE | | BUS. PHONE ( ) | |
| CODE | NAME (Last, First, Middle) | | OCCUPATION | | D.O.B. | AGE | SEX ☐ 1. M ☐ 2. F | RACE ☐ 1. WHT ☐ 2. HISP ☐ 3. BLK ☐ 5. CHI ☐ 7. FIL ☐ 9. P.ISL. ☐ 4. IND ☐ 6. JAP ☐ 8. OTH.____ |
| RESIDENCE ADDRESS | | | | CITY | ZIP CODE | | RES. PHONE ( ) | |
| BUSINESS ADDRESS | | | | CITY | ZIP CODE | | BUS. PHONE ( ) | |

**VIC VEH**

| LICENSE # | STATE | YEAR | MAKE | MODEL | BODY STYLE ☐ 0 UNK ☐ 2 4-DR ☐ 4 P/U ☐ 6 VAN ☐ 8 RV ☐ 10 OTHER ___ ☐ 1 2-DR ☐ 3 CONV ☐ 5 TRUCK ☐ 7 S/W ☐ 9 M/C |
|---|---|---|---|---|---|
| COLOR/COLOR | OTHER CHARACTERISTICS (i.e. T/C Damage, Unique Marks or Paint, etc.) | | | DISPOSITION OF VEHICLE | |

**FACTORS**

- ☐ 1 THERE IS A WITNESS TO THE CRIME        SUSPECT PAGE ☐ YES ☐ NO
- ☐ 2 A SUSPECT WAS ARRESTED
- ☐ 3 A SUSPECT WAS NAMED
- ☐ 4 A SUSPECT CAN BE LOCATED
- ☐ 5 A SUSPECT CAN BE DESCRIBED
- ☐ 6 A SUSPECT CAN BE IDENTIFIED
- ☐ 7 A SUSPECT VEHICLE CAN BE IDENTIFIED
- ☐ 8 THERE IS IDENTIFIABLE STOLEN PROPERTY
- ☐ 9 THERE IS A SIGNIFIANT M.O.
- ☐ 10 SIGNIFICANT PHYSICAL EVIDENCE IS PRESENT
- ☐ 11 THERE IS A MAJOH INJURY/SEX CRIME INVOLVED
- ☐ 12 THERE IS A GOOD POSSIBILITY OF A SOLUTION
- ☐ 13 FURTHER INVESTIGATION NEEDED
- ☐ 14 CRIME IS GANG RELATED
- ☐ 15 HATE CRIME RELATED

**EVIDENCE**

- ☐ 0 NONE
- ☐ 1 FINGERPRINTS
- ☐ 2 TOOLS
- ☐ 3 TOOL MARKINGS
- ☐ 4 GLASS
- ☐ 5 PAINT
- ☐ 6 BULLET CASING
- ☐ 7 BULLET
- ☐ 8 RAPE KIT
- ☐ 9 SEMEN
- ☐ 10 BLOOD
- ☐ 11 URINE
- ☐ 12 HAIR
- ☐ 13 FIREARMS
- ☐ 14 PHOTOGRAPHS
- ☐ 15 OTHER (DESCRIBE)

| VICTIMS SIGNATURE | DATE | DETECTIVE ASSIGNED SIGNATURE | DATE |
|---|---|---|---|
| | | | |

| REPORTING OFFICER | ID# | DATE | REVIEWING SUPERVISOR | ID# | DATE |
|---|---|---|---|---|---|
| | | | | | |

| COPIES: | ☐ CHIEF ☐ CII ☐ PATROL ☐ DB ☐ OTHER | ROUTED BY | ENTERED BY |
|---|---|---|---|
| TO: | ☐ DMV ☐ CAU ☐ ABC (2 copies) ☐ DA ___ | | |

ESPD Form 3/97

**Figure 5-1**    (Courtesy of the El Segundo Police Department)

| | CASE NO. | | PAGE |
|---|---|---|---|

| PREMISES | 7 | POINT OF ENTRY | 8 | PROPERTY ATTACKED | 11 | SUSPECT(S) ACTIONS | 14 | SUSP. PRET. TO BE | 15 |
|---|---|---|---|---|---|---|---|---|---|

**PREMISES** 7 — Q03

**BUSINESS**
- ☐ 1 Bank/Sav Loan Finance/Credit Un
- ☐ 2 Bar
- ☐ 3 Cleaners/Laundry
- ☐ 4 Construction Site
- ☐ 5 Theater
- ☐ 6 Fast Foods
- ☐ 7 Gas Station
- ☐ 8 Hotel/Motel
- ☐ 9 Dept./Discount Store
- ☐ 10 Fast Foods
- ☐ 11 Gun/Sport Goods
- ☐ 12 Jewelry Store
- ☐ 13 Liquor Store
- ☐ 14 Photo Stand
- ☐ 15 Convenience Store
- ☐ 16 Resturant
- ☐ 17 Supermarket
- ☐ 18 TV/Radio
- ☐ 19 Auto Parts
- ☐ 20 Bicycle Sales
- ☐ 21 Car/Motorcycle Sales
- ☐ 22 Clothing Store
- ☐ 23 Hardware
- ☐ 24 Medical
- ☐ 25 Office Building
- ☐ 26 Shoe Store
- ☐ 27 Warehouse
- ☐ 28 Other _____

**RESIDENCE**
- ☐ 29 Apartment
- ☐ 30 Condominium
- ☐ 31 Duplex/Fourplex
- ☐ 32 Garage Attached
- ☐ 33 Garage Detached
- ☐ 34 House
- ☐ 35 Mobile Home
- ☐ 36 Other _____

**PUBLIC**
- ☐ 37 Church
- ☐ 38 Hospital
- ☐ 39 Park/Playground
- ☐ 40 Parking Lot
- ☐ 41 Public Building
- ☐ 42 School
- ☐ 43 Shopping Mall
- ☐ 44 Street/Hwy/Alley
- ☐ 45 Other _____

**VEHICLE**
- ☐ 46 Camper
- ☐ 47 Motor Home
- ☐ 48 Passenger Car
- ☐ 49 Pick-up
- ☐ 50 Trailer
- ☐ 51 Truck
- ☐ 52 Van
- ☐ 53 Other _____

**POINT OF ENTRY** 8 — Q04
- ☐ 0 Unknown
- ☐ 0 N/A
- ☐ 1 Front
- ☐ 2 Rear
- ☐ 3 Side
- ☐ 4 Door
- ☐ 5 Window
- ☐ 6 Sliding Glass Door
- ☐ 7 Basement
- ☐ 8 Roof
- ☐ 9 Floor
- ☐ 10 Wall
- ☐ 11 Duct/Vent
- ☐ 12 Garage
- ☐ 13 Adj. Building
- ☐ 14 Ground Level
- ☐ 15 Upper Level
- ☐ 16 Other _____

**METHOD OF ENTRY** 9 — Q05
- ☐ 0 Unknown
- ☐ 0 N/A
- ☐ 1 No Force Used
- ☐ 2 Attempt Only
- ☐ 3 Bodily Force
- ☐ 4 Bolt Cut/Pilers
- ☐ 5 Channel Lock/Pipe Wrench/Vice Grips
- ☐ 6 Saw/Drill/Burn
- ☐ 7 Screwdriver
- ☐ 8 Tire Iron
- ☐ 9 Unk Pry Bar
- ☐ 10 Coat Hanger Wire
- ☐ 11 Key Slip Shim
- ☐ 12 Punch
- ☐ 13 Remove Louvers
- ☐ 14 Window Smash
- ☐ 15 Brick/Rock
- ☐ 16 Hid inBuilding
- ☐ 17 Other _____

**VEHICLE ENTRY** 10 — Q05
- ☐ 0 Unknown
- ☐ 0 N/A
- ☐ 1 Door/Lock Forced
- ☐ 2 Trunk Forced
- ☐ 3 Window Broken _____
- ☐ 4 Window Forced _____
- ☐ 5 Window Open _____
- ☐ 6 Unlocked
- ☐ 7 Other _____

**PROPERTY ATTACKED** 11 — Q07
- ☐ 0 Unknown
- ☐ 0 N/A
- ☐ 1 Cash Notes
- ☐ 2 Clothes/Fur
- ☐ 3 Consumable Goods
- ☐ 4 Firearms
- ☐ 5 Household Goods
- ☐ 6 Jewelry Metals
- ☐ 7 Livestock
- ☐ 8 Office Equipment
- ☐ 9 TV/Radio/Camera
- ☐ 10 Miscellaneous
- ☐ 11 Other _____

**SEX CRIMES ONLY** 12
- ☐ 1 Suspect Climaxed
- ☐ 2 Unknown/Climaxed
- ☐ 3 Victim Bound/Tied
- ☐ 4 Victim Injured
- ☐ 5 Covered Victim Face
- ☐ 6 Photographed Victim
- ☐ 7 Vic Orally Coupulated Susp
- ☐ 8 Susp OrallyCopulated Vic
- ☐ 9 Rape By Instrument (Foreign Objects)
- ☐ 10 Sodomy
- ☐ 11 Suggested Vic Commit Lewd Perverted Act
- ☐ 12 Inserted Finger into Vagina
- ☐ 13 Forced Vic to Fondle Susp
- ☐ 14 Susp Fondled Victim
- ☐ 15 Masturbated Self
- ☐ 16 Other _____

**BURGLARY ONLY** 13 — Q09
- Is member of Neigh Watch?
  ☐ YES  ☐ NO
- Is member of Operation Ident
  ☐ YES  ☐ NO
- Interested in NW?
  ☐ YES  ☐ NO
- Had Home Business Inspection
  ☐ YES  ☐ NO
- When? _____

**SUSPECT(S) ACTIONS** 14 — Q10
- ☐ 1 Alarm Dismarmed
- ☐ 2 Arson
- ☐ 3 Ate/Drank on Premises
- ☐ 4 Blindfolded Victim Bound/Gagged
- ☐ 5 Cat Burglar
- ☐ 6 Defecated/Urinated
- ☐ 7 Demanded Money
- ☐ 8 Disrobed Victim Fully
- ☐ 9 Disrobed Victim Partially
- ☐ 10 Fired Weapon
- ☐ 11 Forced Vic to Move
- ☐ 12 Forced Vic into Veh
- ☐ 13 Has Been Drinking
- ☐ 14 Indication Multi susps.
- ☐ 15 Inflicted injuries
- ☐ 16 Knew Location of Hidden Cash
- ☐ 17 Made Threats
- ☐ 18 Placed Property in Sack/Pocket
- ☐ 19 Prepared Exit
- ☐ 20 Ransacked
- ☐ 21 Ripped/Cut Clothing
- ☐ 22 Selective in Loot
- ☐ 23 Shut Off Power
- ☐ 24 Smoked on Premises
- ☐ 25 Searched Victim
- ☐ 26 Struck Victim
- ☐ 27 Susp Armed
- ☐ 28 Threatened Retailation
- ☐ 29 Took Only Consumables
- ☐ 30 Took Victim's Vehicle
- ☐ 31 Tortured
- ☐ 32 Under Influence Drugs
- ☐ 33 Used Demand Note
- ☐ 34 Used Lockout
- ☐ 35 Used Driver
- ☐ 36 Used Match/Candle
- ☐ 37 Used Victim Name
- ☐ 38 Used Victim's Suitcase/Pillowcase
- ☐ 39 Used Victim's Tools]
- ☐ 40 Veh Needed to Remove Property
- ☐ 41 Cut/Disconnected Phone
- ☐ 42 Cased Location Before Crime
- ☐ 44 Other _____

**SUSP. PRET. TO BE** 15 — Q11
- ☐ 0 N/A
- ☐ 1 Conducting Survey
- ☐ 2 Cust./Client
- ☐ 3 Delivery Person
- ☐ 4 Disabled Motorist
- ☐ 5 Drunk
- ☐ 6 Employee/Employer
- ☐ 7 Friend/Relative
- ☐ 8 Ill/Injured
- ☐ 9 Need Phone
- ☐ 10 Police/Law
- ☐ 11 Renter
- ☐ 12 Repairman
- ☐ 13 Sale of Illicit Goods
- ☐ 14 Sales Serson
- ☐ 15 Seek Assistance
- ☐ 16 Seek Directions
- ☐ 17 Seeking Someone
- ☐ 18 Solicit Funds
- ☐ 19 Other _____

**PHYSICAL SECURITY** 16 — Q12
- ☐ 0 Unknown
- ☐ 0 N/A
- ☐ 1 Audible Alarm
- ☐ 2 Slient Alarm
- ☐ 3 Private Security Patrol
- ☐ 4 Dog
- ☐ 5 Standard Locks
- ☐ 6 Auxiliary Locks (Deadbolt Windows, etc.)
- ☐ 7 Window Bars/Grills
- ☐ 8 Outside Lighting On
- ☐ 9 Inside Lighting On
- ☐ 10 Garage Door Locked
- ☐ 11 Obscured Interior View (Commercial/Business)
- ☐ 12 Security Signing (N.W., Alarm, etc.)
- ☐ 13 Other _____

**VICTIM PROFILE**

**PHYSICAL CONDITION** 17 — Q13
- ☐ 0 No Impairment
- ☐ 1 Under Infl. Alcohol/Drugs
- ☐ 2 Sick/Injured
- ☐ 3 Senior Citizen
- ☐ 4 Blind
- ☐ 5 Handicapped
- ☐ 6 Deaf
- ☐ 7 Mute
- ☐ 8 Mentally/Emotionally Impaired
- ☐ 9 Other _____

**RELATIONSHIP TO SUSPECT** 18 — Q14
- ☐ 0 Unknown
- ☐ 1 Husband
- ☐ 2 Wife
- ☐ 3 Mother
- ☐ 4 Father
- ☐ 5 Daughter
- ☐ 6 Son
- ☐ 7 Brother
- ☐ 8 Sister
- ☐ 9 Other Family
- ☐ 10 Acquaintance
- ☐ 11 Friend
- ☐ 12 Boyfriend
- ☐ 13 Girlfriend
- ☐ 14 Neighbor
- ☐ 15 Business Associate
- ☐ 16 Stranger
- ☐ 17 Other _____

**MARTIAL STATUS** 19 — Q15
- ☐ 0 Unkown
- ☐ 1 Annulled
- ☐ 2 Common Law
- ☐ 3 Single
- ☐ 4 Married
- ☐ 5 Divorced
- ☐ 6 Window(er)
- ☐ 7 Separated
- ☐ 8 Other _____

**Figure 5-2**   (Courtesy of the El Segundo Police Department)

report face sheets there are a number of similarities in the type of information they ask for. Some of these commonly requested items include:

**COPIES TO.** This will generally be located in one of the corners of the face sheet and should be used to write the name of any investigator within the agency or the name of any outside agency that the writer wishes to receive a copy of the report.

**CASE NUMBER.** This refers to desk report and crime report numbers. These terms describe numbering systems used to file and index the documents and reports an investigative agency is involved with.

**PREPARED BY.** This is asking for the name, rank, and any badge or identification number used by the preparer.

**OCCURRED ON.** The date and time of occurrence and the day of the week is entered in this area. If the exact date or time cannot be established, the dates or time within which the crime took place should be listed.

**CRIME.** In the space calling for the crime, the investigator should place the code section and name of the crime. Lesser and included offenses should not be listed when perpetrated against a single victim. Generally, the most serious crime shown by the elements present is used and only crimes substantiated in the narrative section of the report should be listed.

**DATE AND TIME REPORTED.** In this section write the date and time the crime was first reported to your agency. This could be the time a complaint operator received the call, or when you were first spoken to by a victim or complainant in the field.

**RD or REPORTING DISTRICT.** Enter the reporting district number where the crime occurred. Several reporting districts usually comprise an area or beat, and are used to determine activity levels and calls for service in a geographical space.

**LOCATION OF CRIME.** In the block asking for the location of the occurrence or crime, pinpoint the location as closely as possible. In a report where a business is listed as the victim, it is imperative that the business address be recorded. When the specific location is impossible to determine, use the vicinity of the crime. If the location is the same as that of the victim's address you can generally write SAME AS ABOVE, as long as it is clear what you are referring to.

**CSI.** There is usually a box to check if a crime scene investigator or evidence technician examined the crime scene. It may also be necessary to write their name and identification number in the report.

**SOLVABILITY FACTORS.** Sometimes referred to as FACTORS, these are general questions that the investigator can answer by checking a yes or no box for each question asked. Solvability factors are used by supervisors and follow-up investigators to prioritize those cases with the greatest chance of solution based on the information and

**EL SEGUNDO POLICE DEPARTMENT**   **ADDITIONAL VICTIMS/WITNESSES**

PAGE_____ OF _____

CASE NO

| CRIME 1 | CODE SECTION | CRIME | CLASSIFICATION | REFER OTHER REPORTS |
|---|---|---|---|---|
| | LOCATION | | RD. | DATE | TIME | SUPPL. ☐ | INCIDENT NO. |

**WITNESS/RP/OR ADD. VICTIM** — 2

For each entry block:

| CODE | NAME (Last, First, Middle) | OCCUPATION | D.O.B. | AGE | SEX | RACE |
|---|---|---|---|---|---|---|
| | | | | | ☐ 1. M  ☐ 2. F | ☐ 1. WHT  ☐ 2. HISP  ☐ 3. BLK  ☐ 5. CHI  ☐ 7. FIL  ☐ 9. P.ISL.  ☐ 4. IND  ☐ 6. JAP  ☐ 8. OTH._____ |

| RESIDENCE ADDRESS | CITY | ZIP CODE | RES. PHONE ( ) |
|---|---|---|---|

| BUSINESS ADDRESS | CITY | ZIP CODE | BUS. PHONE ( ) |
|---|---|---|---|

(This entry block is repeated eight times.)

| REPORTING OFFICER | ID# | DATE | REVIEWED BY | ID# | DATE |
|---|---|---|---|---|---|

COPIES: ☐ CHIEF  ☐ CII  ☐ PATROL  ☐ DB  ☐ OTHER AGENCY   ROUTED BY    ENTERED BY
TO: ☐ DMV  ☐ CAU  ☐ ABC (2 copies)  ☐ DA  _____

ESPD Form #269 (Rev 5/97)

**Figure 5-3**   (Courtesy of the El Segundo Police Department)

evidence available. There are no hard and fast rules for selecting a positive or negative answer, and no single answer is a determining factor in the decision to continue the investigation.

**VICTIM'S NAME.** In this section, write the last name first, then the first name, followed by the full middle name last. The last name might be followed by a comma or underscored, or both when the name could be mistaken for a given name. If the report is to be typed, the last name might be in upper case type to help distinguish it. When the victim has no middle name or initial, write NMN, which means No Middle Name. If the victim is a company, the company name should be written. If an ABC Oil Company service station is burglarized and money belonging to the company is taken, the victim is the ABC Oil Company. If a robbery is committed against the manager or any employee of the same service station in which company money or property is taken, the firm is the victim and the employee would be a witness. If in addition to company money, the personal funds and property of an employee are taken, the employee would also be named as a victim. If more than one victim is involved in the same offense, the symbol V-1 should precede the name written in this section, and the names of other victims should be preceded by the symbols V-2, V-3, and so on. Additional victims should be listed in the narrative section and appropriately titled.

**VICTIM'S ADDRESS.** In this section, list the principal victim's address when the victim is an individual. If the victim is a business, list the business address.

**RESIDENCE PHONE.** In the box asking for the residence phone, list the home phone number of the victim, or in the case of a business, the residence phone of the owner. *Always* precede the number with the area code.

**BUSINESS PHONE.** In this section write the telephone number at which the victim can be reached while at work. *Always* precede the number with the area code. If the victim is unemployed, write NONE. In the event the victim is a place of business, write the telephone number of the business, and write DAY in front of the daytime phone number.

**VICTIM'S OCCUPATION.** In this box write the victim's normal means of making a living. If the victim is unemployed, use the type of occupation the victim is normally employed in and write, for example, UNEMPLOYED CARPENTER. When the victim is a business establishment, write the type of business that is conducted there, unless the name of the business precludes the necessity of further explanation, for example, JOE'S BAR. If the victim is a student, write the name of the school attended and the city where the school is located (for example, Student, City High School, San Diego).

**DOB.** Write the victim's date of birth when available and applicable.

**PERSON REPORTING OFFENSE.** In many instances, this will be the same individual previously listed as the victim. If such is the case, write the victim's last and first names here. However, if the person reporting the offense is someone other than the victim, this should be recorded. If the victim is a business and the reporting party is an employee, his relationship with the business should be described.

**PERSON WHO DISCOVERED THE CRIME.** Many times this will be the same person as the victim or the person reporting. If so, the details for this box have already been obtained. If, however, the person reporting the offense is someone other than those mentioned, this should be recorded. If the victim is a business and the person discovering the crime is an employee, this relationship with the business should be shown.

**MO SECTION.** The MO or Modus Operandi section is generally completed for robberies, burglaries, and aggravated sex cases, but may also be completed on other types of reports when the investigator believes the information would be helpful in solving the crime. This is the area to write the basic method used by the suspect to commit the crime, along with any unusual things the suspect did during its commission.

**CHARACTERISTICS OF PREMISES OR AREA.** This is where the characteristics or type of place where the crime was committed are written. This information, in many cases, will be a description of the size, type, area, or characteristics of the neighborhood. Examples could be, "One story, five room, single family residence on a corner lot," or "Mobile home in a largely unattended mobile home park."

**MOTIVE-TYPE OF PROPERTY TAKEN OR OTHER REASON FOR OFFENSE.** Generally, the class or type of property taken, or the motive or reason why the offense was committed is written in this area. In crimes where property has been taken, the motive will usually be the type of property taken in support of personal gain. The specific type of property that was taken or was attempted to be taken should be listed. This might include money and jewelry, women's clothing, cigarettes, or narcotics. The detailed description of the property and any serial numbers should appear in the Property of Loss Section. In other types of crimes, the motive might be revenge, insurance settlement, concealment of crime, sexual gratification, ransom, or in narcotic cases money from the sale of narcotics or the effects resulting from their use. In some cases, the initial crime may lead to a second offense, for example, a case in which a homicide is committed during a robbery or attempted rape. In such cases the motive would be robbery or rape.

**VICTIM'S ACTIVITY JUST PRIOR TO AND/OR DURING THE OFFENSE.** The victim's activity at these times may characterize the kind of person the offender selected as a victim. In rape cases, the victim's activity just prior to the attack might be "waiting at bus stop," "doing laundry in a laundromat," "entering car in parking lot," or "in bed asleep." In a robbery case, the victim's activity just prior to the offense might be "walking down street," "waiting on customers," or "closing store." With a burglary case, the victim's activity during the offense might be "on vacation," "attending a funeral," or "home in bed." When the victim is a business, the victim's activity is likely to be either "open for business" or "closed for business." When open for business is used, include the natural activity of the attendant or employee just prior to or during the attack.

**DESCRIBE WEAPON, INSTRUMENT, EQUIPMENT, TRICK, DEVICE, OR FORCE USED.** For crimes against property, list the type of tool used and its size if this can be determined. If tools are not used, write what was used such as, hands, feet, voice, etc. For crimes against persons, write as complete a description of the weapons used. If force was used, describe the force; for example, "knocked to ground," "kicked," "hit with fist," or "threatened with unknown type liquid."

**WHAT DID SUSPECT SAY.** If possible write the exact words used by the suspect. Pay particular attention to any mispronunciations, unusual words or peculiar expressions, accents, or dialects. Many times how the suspect said something is just as important as what was said. Try to explain this if it is at all possible.

**TRADEMARK, OTHER DISTINCTIVE ACTION OF SUSPECT.** Any action by the suspect in preparation for the crime, flight from the scene, or disposition of the proceeds of the crime that has not been recorded in any other category of the modus operandi should be written here. The act may be necessary for the completion of the crime, but frequently is not. Preparations for the crime, as well as precautions to avoid apprehension or detection, may be necessary but are not included elsewhere in the report. Examples are "cased store room the day before," "wiped off fingerprints," or "closed venetian blinds but turned one slat to provide view of front entrance." Unnecessary acts are "eats food," "leaves note," or "plays stereo." The number of bizarre acts a suspect may do are unlimited.

**ADDITIONAL VICTIMS.** List additional victims and include all of the information you did for victim #1 such as name, address, residence phone, business phone, occupation, and date of birth. Identify additional victims as V-2, V-3, etc.

**WITNESSES/SUSPECTS.** When there is more than one witness or suspect, number them using the same guidelines as for additional victims. Include as much information and as complete a description as possible. When a suspect is in custody, say so by writing IN CUSTODY followed by the name of the jail. Generally, suspects should be listed by name and description only in cases when they could be arrested or a complaint charging them with the offense could be issued. Otherwise, list them in the narrative section of the report. Suspects' descriptions should be written separately in the event different witnesses give different descriptions. If something is not known about a person's description, leave it out, and *never* combine suspect descriptions into one.

**VEHICLE USED BY SUSPECTS.** Write any information available as to the transportation used by the suspect. This may range from a complete description of an automobile to a partial one based on the size and number of tire tracks seen at the crime scene. When you cannot establish that a suspect used a vehicle, write NONE SEEN OR HEARD.

**PROPERTY.** Each item should be numbered for easy indexing and the quantity of each type of item stolen should also be shown. Following this, write a complete description of the property including the make, model, serial number, size, value, and any other identifiable marks or characteristics. Remember the AVERAGE PERSON TEST. A total value for the stolen property is usually called for in this area as well.

**EVIDENCE.** If evidence is collected by the investigating officer or someone other than a crime scene investigator, the person responsible for the chain of custody should be listed on the report in this area. If a crime scene investigator collects the evidence he or she will most likely prepare a report using the same case number as your investigation for easy reference. If this is the case, write SEE CRIME SCENE INVESTIGATOR REPORT.

**INJURIES.** Describe who received injuries and the seriousness of them.

Your reputation may rest on a report you write. Many people will read and evaluate your work and you may have to base your testimony on your report as well. There may be times when information is limited, but you must make every reasonable attempt to complete the report. It may be possible that the best opportunity for solving a crime rests with the first investigator at the scene; when the evidence is fresh and witnesses are more likely to be present and remember what happened. Some consider radio calls of a theft or a burglary as a task in which they have to *take a report*. In reality, each of these calls for service is an opportunity to investigate a crime and perhaps solve it. Regardless of your rank, assignment, or tenure, if you are assigned to conduct a lawful search for a thing or person, and the goal is to find the truth, you are an investigator.

## REVIEW

1. Crime reports must establish the corpus dilecti.
2. The purpose of a crime report face sheet is to organize information and gather statistics.
3. The two parts to a crime report are a fill-in-the-blanks face sheet and a free-flowing narrative.
4. Start the narrative with the date, time, and how you got involved.
5. Solvability factors help prioritize follow-up investigation.

**EXERCISES**

1.  Visit three police departments in your area and get copies of the crime report face sheets used by them. What are the similarities?
2.  Are there any significant differences among the reports? How will this affect the preliminary investigation of an incident?
3.  View a law enforcement action show and take notes on the calls for service. Using a face sheet, complete a crime report.
4.  Divide the class into two groups with one group playing the victim and the other the officer. Using scenarios supplied by the instructor, conduct a preliminary investigation and complete a crime report.
5.  Review the crime report and identify the areas needing improvement.
6.  Rewrite the crime report using the rules of narrative writing. Any differences in the final product?
7.  Crime Report Exercise. Review the crime report and identify any problems. Rewrite the report using the rules of narrative writing.

---

CRIME:   Vandalism        RD: 637    BEAT: 2
LOCATION:    187 Van Ness Circle, El Fuego, CA
OCCURRENCE DATE:    7-6-99 to 7-7-99 between 2300-0530
REPORTED DATE:    7-16-99    1625 hours
REPORTING PERSON:   Cleworth, Frank    DOB 12-3-34
                    187 Van Ness Cr. #19C
                    El Fuego, CA
                    Home Phone (800) 555-6826
                    Work Phone (800) 555-6826
DISCOVERER:    Same as above
VICTIM:    Clear Picture Cable
           154 Roswell, El Fuego (800) 555-4910

---

THERE IS NO M.O. SECTION FOR THIS REPORT

---

SOLVABILITY FACTORS
     [Y]    IS THERE A SUSPECT?
     [Y]    IS THIS A CONTINUING PROBLEM?

---

PROPERTY DAMAGE:    2-1/2′ of television cable, nfd; value $50.
DETAILS:

---

At the above indicated date and time, I was dispatched to the above location. Upon my arrival at the Van Ness address, I contacted the R/P, Mr. Cleworth.

Mr. Cleworth stated to me that the gentleman who lives in the apartment above him, Mr. Paul Johnson (unknown spelling), who resides at 187 Van Ness, apt. #19G, has been an ongoing dispute and argument with the R/P.

He stated that on Tuesday evening, 7-6-99, at approximately 2300 hours, he terminated watching television for the evening and retired. The very next morning, on Wednesday, 7-7-99, at approximately 0530 hours, he awoke and responded to the living room of his condominium. There, he turned on the television, which was functioning, but was unable to receive any picture. He immediately contacted the Clear Picture Cable people that same day. Mr. Cleworth indicated that Clear Picture Cable driver Paul V. S., #27 responded to his residence. There, he checked the cable.

It should be noted that there is an ordinance in the C.C.R.'s of the condominium complex that requires all cables to either be underground or concealed within the building. As a result, the television cable system that Mr. Cleworth ordered had to be installed through the attic of the two story condominium building. From the attic, the cable was then run down along the inside of the exterior laundry room attached to the balcony of apartment #19G belonging to Mr. Johnson; there, through the floor, into the laundry room located on the ground level below Mr. Johnson's condo belonging to Mr. Cleworth, and then from the laundry room, into the living room of Mr. Cleworth's condominium located directly below that of Mr. Johnson.

Mr. Cleworth indicated that the cable television installer was unable to locate any difficulty with the cable on the ground level and that he was unable to locate any difficulty with the cable leading to the building, but was unable to gain access to Mr. Johnson's condominium at that time.

The same Clear Picture Cable representative returned to the location on 7-10-99. There, with the consent of the manager of the condominium complex, Mr. Joe Willy (unknown residence number, business phone number: 555-5475); the Clear Picture Cable representative, Paul V. S., #27, and the manager, Mr. Willy, climbed the ladder onto the balcony of condominium #19G belonging to Mr. Johnson. There, they opened the door leading to the laundry room located on the balcony and examined the interior. There, inside Mr. Johnson's laundry room, the installer discovered a 2-1/2' piece of the television cable to be cut away from the cable leading from the attic above, down to Mr. Cleworth's condominium. The installer subsequently repaired the splice, advised Mr. Cleworth of what he had found, and left the location.

This reporting officer did not see the damage indicated by Mr. Cleworth. Mr. Cleworth also indicated that the repairs were at the expense of the victim, Clear Picture Cable, who was at this time unavailable for comment.

Mr. Willy, the manager of the complex, was also unavailable for comment.

## QUIZ

1. When should a crime report be completed?
2. What is the most important thing to be established in a crime report?
3. What is the most important use of a crime report?
4. What are the two basic parts of a crime report?
   a.
   b.
5. Why is the phrase "take a report" a misnomer?
6. What are solvability factors?
7. What is the purpose of a crime report face sheet?
8. In what time frame should a crime report be completed?
9. What is the minimum length of a crime report narrative?
10. How should the narrative of a crime report begin?

C H A P T E R

# 6 Arrest Reports

Generally speaking, an arrest report is necessary whenever someone is taken into custody by a law enforcement agency as the result of a criminal investigation, or in response to a citizen's request for a private person's arrest. Although fact patterns indicating an arrest are infinite, there are four basic reasons an arrest is made.

1. A police officer sees a misdemeanor crime occur.
2. A police officer believes a felony has been committed by a particular person.
3. A citizen makes a private person's arrest.
4. An arrest warrant exists for a person.

## DOCUMENTING THE ARREST

Whichever one of these circumstances is the basis for the arrest, the responsibility for documenting it rests with the investigator who detains the suspect. His or her report should clearly show the circumstances of what happened and include the facts that led to the decision to make the arrest. These facts and circumstances combine to form the **probable cause,** which is the basis for the arrest. Without exception, the single most important thing that must be established in an arrest report is probable cause.

### Arrest Report Styles

Depending on the circumstances and the length of the investigation, an arrest report might take one of several forms. Some agencies use an arrest report face sheet that is similar in appearance and function to a crime report face sheet in that it helps to organize information and allows for the easy gathering of statistical information. This FILL-IN-THE-BLANKS form also requires a narrative section to be completed. Other agencies use a total narrative style for arrest reports and gather statistics from the booking sheet.

Whether the combination face sheet and narrative style or the all narrative style is used, the key to preparing a complete, clear, concise, and accurate arrest report is to write a narrative section that explains what happened. You must explain who did what, how it happened, and where and when it took place. The Rules of Narrative Writing described in chapter 3 apply to arrest report narratives as well and allow them to be completed quickly, professionally, and with consistency. The writer of an arrest report will soon realize that the narrative of a crime report and the narrative of an arrest report follow the same rules. They are written in the *first person, past tense, active voice; in chronological order beginning with the date, time, and how you got involved; and use short, clear, concise, and concrete words.*

**LA HABRA POLICE DEPARTMENT**
**DUI FIELD INTERVIEW REPORT**

| CASE NUMBER | | |
|---|---|---|
| PAGE | | OF |

| DRIVER'S LAST NAME | DRIVERS LICENSE NUMBER | STATE | CLASS | STATUS | DATE / TIME OF FIELD INTERVIEW |
|---|---|---|---|---|---|

PASSENGER NAME/DOB/ADDRESS/PHONE

| VEH | YEAR | MAKE | MODEL | STYLE | COLOR | LICENSE PLATE # | STATE | ACCIDENT INVOLVED? ☐ YES   ☐ NO |
|---|---|---|---|---|---|---|---|---|

LOCATION OF :   ☐ TRAFFIC STOP   ☐ TRAFFIC COLLISION    R/O NAME & ADDRESS:

1. Do you know of anything mechanically wrong with your vehicle?   ☐ YES   ☐ NO    If yes, describe: _____

2. Are you sick or injured?   ☐ YES   ☐ NO   If yes, describe: _____

3. What time is it? _____ AM/PM        3a. Actual time? _____ Hrs.

4. Are you diabetic or epileptic?   ☐ YES   ☐ NO   If yes, describe: _____

5. Do you take insulin pills or injections?   ☐ YES   ☐ NO   If yes, describe: _____

6. Do you have any physical defects?   ☐ YES   ☐ NO   If yes, describe: _____

7. When did you last sleep? _____ AM/PM        7a. How long? _____

8. When did you last eat? _____ 8a. Describe meal: _____

9. Have you bumped your head recently?   ☐ YES   ☐ NO   If yes, describe: _____

10. (If driving not observed)  Were you driving the vehicle?   ☐ YES   ☐ NO

11. How long have you been driving today/tonight? _____

12. Where did you start driving? _____

13. Where were you going? _____

14. Where are you now? _____

15. What have you been drinking? _____

16. How much have you been drinking? _____

17. What time did you start drinking? _____

18. What time did you stop drinking? _____

19. Where were you drinking? _____

20. With whom were you drinking? _____

21. Do you feel any effects from the drinks?   ☐ YES   ☐ NO   If yes, describe: _____

22. Are you currently under the care of a Doctor or Dentist?   ☐ YES   ☐ NO   If yes, name and adress: _____

23. Have you had any recent surgery?   ☐ YES   ☐ NO   If yes, describe: _____

24. Have you taken any medicine?   ☐ YES   ☐ NO   If yes, describe: _____

25. What dosage? _____

26. What was the time of the last dose taken? _____

27. Have you had anything to drink in the last hour?   ☐ YES   ☐ NO   If yes, describe: _____

DESCRIBE TEST LOCATION:  (WEATHER CONDITIONS, LIGHTING, GRADE ETC)

**INVESTIGATION NOTES**

| REPORTING OFFICER / ID | WITNESSING OFFICER / ID | APPROVED BY |
|---|---|---|

**Figure 6-1**   (Courtesy of the La Habra Police Department)

**LA HABRA POLICE DEPARTMENT**
**DUI FIELD BALANCE TEST**

CASE NUMBER:

DATE/TIME OF F.B.T.

| BREATH | EYES | SPEECH | COORDINATION | APPEARANCE | BEHAVIOR | CLOTHING |
|---|---|---|---|---|---|---|
| ALCOHOLIC MARIJUANA OTHER/PCP OTHER____ _____ STRONG MODERATE WEAK NONE | BLOODSHOT WATERY DROOPY CLEAR PUPIL SIZE: CONSTRICT DILATED OTHER____ | SLURRED LOUD SOFT/QUIET MUMBLED SPITTING RAPID TALKATIVE SURE/ CORRECT OTHER___ | UNSTEADY/ GAIT USED/ SUPPORT STAGGERED FELL FUMBLED ID LAX FACE/ JAW OTHER | UNKEMPT VOMITUS HAIR MESSED DRY MOUTH NORMAL OTHER____ | ANGER LAUGHING CRYING BELLIGERENT AGGRESSIVE ARROGANT REMORSEFUL INDIFFERENT FLUCTUATING COOPERATIVE | UNBUTTONED UNZIPPED DISHEVELED URINE NEAT STAINED/ DIRTY DESCRIBE: ____ _____ |

| GLASSES CONTACTS | YES NO | SHOES WORN DURING TEST | YES NO | DESCRIBE SHOES: | NYSTAGMUS TEST GIVEN | YES NO |

## FIELD BALANCE TESTS

| ALPHABET TEST | MODIFIED POSITION OF ATTENTION | ONE LEG STANCE |
|---|---|---|
| SURE, CORRECT SLOW, HESITANT SLOW, DELIBERATE STARTED OVER____TIMES SLURRED NOT UNDERSTANDABLE | SURE HESITANT USED ARMS FOR BALANCE LOST BALANCE FALLING OTHER_____ LATERAL SWAY_____" | LEFT — SURE HESITANT USED ARMS FOR BALANCE LOST BALANCE FALLING OTHER — RIGHT |
| SKIPPING LETTERS | EXPLAIN: | EXPLAIN: |
| ABCDEFGHIJKLMNOPQRSTUVWXYZ ABCDEFGHIJKLMNOPQRSTUVWXYZ | | |

### FINGER TO NOSE TEST

X = RIGHT    0 = LEFT

LEFT    SURE HESITANT SLOW/SEARCHING    RIGHT

TURNING HEAD TO MEET FINGER

EXPLAIN:

### HEEL TO TOE WALK

X = RIGHT    0 = LEFT

EXPLAIN:

| ABILITY TO READ AND FOLLOW SIMPLE DIRECTIONS | ( ) GOOD RETENTION & QUICK RESPONSE TO INSTRUCTIONS ( ) POOR RETENTION AND RESPONSE ( ) ATTEMPTS TESTS BEFORE / DURING INSTRUCTIONS | ( ) INTERRUPTING ( ) EVASIVE QUESTIONS ( ) FAIR RETENTION AND RESPONSE |

| 13353 CVC | 13353 CVC ADVISAL BY OFFICER ____ (NAME/ID) ( ) ENGLISH    ( ) REFUSAL ( ) SPANISH | BLOOD TEST SAMPLE TIME:_____HRS. TEST LOCATION:_____ LVN NAME:_____ VIAL #:_____ | BREATH TEST OPERATOR NAME _____ TEST LOCATION _____ RESULT #1 _____% BAC RESULT #2 _____% BAC | URINE TEST SAMPLE TIMES_____HRS _____HRS TEST LOCATION _____ SAMPLE ID# _____ OFFICER NAME / ID _____ |

| REPORTING OFFICER/ / ID | ASSISTING OFFICER / ID | APPROVED BY / DATE |

LHPD 91-11    REVISED 8/91

**Figure 6-2**    (Courtesy of the La Habra Police Department)

| EL   SEGUNDO   POLICE   DEPARTMENT | | CASE NO. | PAGE |
|---|---|---|---|

**SUPPLEMENTAL/NARRATIVE**
*(Check one)*

☐ SUPPLEMENTAL REPORT   ☐ NARRATIVE CONTINUATION   **CA0192300**

| CODE SECTION | CRIME | VICTIM'S NAME   (FIRM IF BUSINESS) |
|---|---|---|

| PREPARED BY | | | | REVIEWED BY | | |
|---|---|---|---|---|---|---|
| NAME | I.D. NUMBER | MO. | DAY | YR. | NAME | MO. | DAY | YR. |

P-16 REV 7-98

**Figure 6-3**   (Courtesy of the El Segundo Police Department)

## Completing the Report

When completing an arrest report that uses a face sheet, begin by filling in the boxes as requested. Then, begin the narrative with the date, time, and how you got involved, for example:

> On 9-1-98 at 2145 hours I received a radio call of a prowler at 6101 Pine Street.
> On 7-4-99 at about 1330 hours I saw Brown, who was driving west on Pacific, fail to stop for the stop sign at Main Street.
> On 12-19-98 at 1720 hours I was driving through the Del Lago Plaza parking lot when Brown haled me and said her husband was chasing a purse snatcher.

When no face sheet is used, the arresting officer must overcome the problems of creating a workable format and beginning the report. One format that works well is to list pertinent headings of ARRESTED, CHARGE, LOCATION, DATE AND TIME, OFFICER, and DETAILS at the top left side of the page, complete the information based on the fact pattern, and begin the narrative. These headings, when set up properly, allow for multiple suspects to be listed while still organizing the format in a workable way. For example:

### One Suspect

ARRESTED:
CHARGE:
LOCATION:
DATE AND TIME:
OFFICER:
DETAILS:

### Two or More Suspects

ARRESTED:        1.
CHARGE:

                 2.

                 3.
LOCATION:
DATE AND TIME:
OFFICER:
DETAILS:

In either case, you would begin the narrative below the heading DETAILS with the date, time, and how you got involved.

Questions about how much detail is needed and what kind of information to include in an arrest report are frequently asked. Ideally, every arrest report would include everything that happened, answer every question anyone reading it might have, and allow for a quick resolution to the matter because of its completeness. Realistically, this is not likely in many cases, but there are some things that are necessary in every case:

```
                    EL  FUEGO  POLICE  DEPARTMENT

ARREST REPORT                                98-6562110


ARRESTED:       Rabbit, Donald DOB 6-18-52

CHARGE:         Warrant F246381

                211 PC, Robbery

LOCATION:       Main and Walnut, El Fuego

DATE and TIME:  12-16-98 1830 hours

OFFICER:        J. Fava #24

DETAILS:

     On 12-16-98 at 1530 hours I received information during

roll call that Rabbit was wanted for a robbery of the Tall

Can Liquor Store at 301 Pacifica Street.  The information

included a physical description and a color photograph.  At

1830 hours I was driving south on Main Street and saw Rabbit

standing in line at the XLT Bus Depot.

     I spoke with him and he told me he was Rabbit.  I

verified the warrant was valid and arrested him.  I booked

Rabbit at the El Fuego City Jail.




J. Fava #24
```

**Figure 6-4**   One-person arrest report format.

1.  Identify all the people present at the location of the arrest, or who participated in it in any way.
2.  Identify and explain all injuries to any person involved, and how they occurred.
3.  Clearly report statements made by the suspect.
4.  Clearly describe the probable cause for the arrest.
5.  If there are multiple charges, make sure each charge is supported with probable cause.

### EL FUEGO POLICE DEPARTMENT

ARREST REPORT                                           99-098418


ARRESTED:        1.    Adams, Paul DOB 11-27-55

CHARGE:                459 PC Burglary

                 2.    Barkton, Steven DOB 3-17-46

                       459 PC Burglary

LOCATION:        423 Raindrop Lane, El Fuego

DATE and TIME:   2-17-99 1500 Hours

OFFICERS:        M. Conklin #199

                 R. Sproul #7654

DETAILS:

On 2-17-99 About 1445 hours we received a radio call of a silent intrusion alarm at 423 Raindrop Lane. We arrived and set up a perimeter and could see Adams and Barkton through the front window. They were standing in front of an entertainment center and were disconnecting wires from the video recorder. Using the car's public address system I announced our presence and ordered them to come out. As they walked out the front door, Sproul arrested them and put them into separate cars.

I spoke to Billy Huntly, 426 Raindrop Lane, (213) 555-1212, and learned that Kathy Czaban (213) 555-1212 is the owner of the house and was away on vacation. I phoned her at her hotel in Los Angeles and learned that Barkton is a former boyfriend who did not have permission to be in the house. There is no loss at this time, but Czaban will make a complete check when she returns.

I booked Barkton and Adams at the El Fuego City Jail.

M. Conklin #199

**Figure 6-5**    Multiple-person arrest report format.

It is not the arresting officer's obligation or duty to prepare a one-sided, biased, or slanted report. This serves no purpose other than to delay the criminal justice process. It is the arresting officer's responsibility to report all the relevant facts in a complete, clear, concise, and accurate manner that will allow all those who read the report to form their own opinions and draw their own conclusions. The circumstances of how you get involved in cases ending with an arrest are many times out of your control. One thing within your control and that you are able to manage is your ability to write a good report that clearly shows who, what, where, when, why, and how you got involved, establishing a factual basis for what you did. The key to writing a good arrest report in any situation is to know the law, know your agency's policies, and do your job within these parameters.

## REVIEW

1. An arrest report is needed whenever someone is taken into custody.
2. The single most important thing to be established in an arrest report is the probable cause.
3. The length of an arrest report is dependent on the facts of the case.
4. Use the Rules of Narrative Writing to complete the narrative of the report.
5. Use short, clear, concise, and concrete words.
6. Limit unsupported opinions.
7. The report must be unbiased.

## EXERCISES

1. Using the following fact pattern, prepare an arrest report using today's date and time.

   FACTS:
   You are a patrol office for the El Fuego Police Department and are on duty as Unit 6. You are on patrol traveling west on Carson Street. As part of your normal and routine patrol duties you drive north on Faculty and as you are approaching the T intersection of Village and Faculty, you see a 1997 red corvette, driving east on Village, come up to the stop sign at Faculty and make a left turn onto Faculty without stopping at the stop sign. The wheels of the corvette never stopped and actually broke traction as the driver accelerated out of the turn. Your best estimate of the car's speed is 5 miles per hour. The license plate on the corvette is a personalized plate of SEEYA. You make a car stop at the next cross street north of the intersection,

which is 400 yards away at the intersection of Faculty and South Street. When you talk to the driver he tells you that he usually does not stop at stop signs if there is no traffic because it is inconvenient. He gives you his operators license and you see that his name is Joseph Paul White and his birth date is July 4, 1963. He also tells you that he never pays his tickets and probably has a warrant out for his arrest. You check his driving record as you write the ticket on Citation number EF9987659 and when you are almost finished you get the following radio message from the dispatcher.

*Dispatch:*    Unit 6
*You:*    Unit 6 go ahead
*Dispatch:*    Unit 6 there is an outstanding burglary warrant for Joseph White, DOB 7-4-63. Bail is $15,000 and the warrant number is F4591015.
*You:*    10-4.

After the driver signs his ticket you arrest him for the warrant and take him to the El Fuego City Jail where you book him. You put his copy of the traffic ticket in his property. Before you took him to jail he asked you to lock up his corvette and leave it where it is parked. The registration in the car showed that the driver is the owner and that he lives at 3007 Raceway Lane, El Fuego.

2.    Using the following fact pattern, prepare an arrest report with an evidence report. Use today's date and time.

FACTS:
You are a patrol officer for the Big Rock Police Department and are working with your partner Officer Ron Ho as Unit 2. As part of your daily briefings, you are familiar with an on-going problem of property damage occurring in the industrial area in which unknown suspects have been spray painting graffiti on the buildings. You have had several conversations with the local business owners about the problem, including Mr. Steven Luck who owns the Good Luck Auto Repair Shop. Mr. Luck has told you several times that if you ever catch someone painting on his property, he will cooperate and support prosecution. While you are patrolling an industrial area you drive around the corner from east bound Container Lane to south bound Cement Avenue and both see a man spray painting the letters BG onto the side of the Good Luck Auto Repair Shop. Your partner jumps out of the car and tells the man to stop, but he finishes the painting. As soon as the painting is done, he throws the can down and starts running away, south on Cement. Officer Ho is in excellent condition and easily catches the suspect after a chase of only 50 yards. The suspect gives up without a struggle and Officer Ho handcuffs him.

While Officer Ho is walking the suspect back to the patrol car, you picked up the spray can. It is a 10 ounce can of red paint made by Brand X. The suspect tells you his name is Trevor Robers and his birthday is February 14, 1965. You and Officer Ho take the suspect to the Big Rock Jail and book him for vandalism.

3.  Review the attached report and identify any issues. Using the arrest report format, the facts provided, and the rules of narrative writing, prepare an arrest report. Use today's date and the time shown in the report.

FACTS:

Today at approx. 1015 a.m. I was taking a theft report at the Brand X gas station at Roscoe & Beach. I heard a broadcast go out that a susp. was in a bank across the street & that he was a possible forger.

I left the report and drove across the St. to the bank. Motor Off. Najjar & Brown assisted. I detained susp. Valley until we got a story from the bank. I ran a check on Valley & 1 outstanding warrant came back to his name. His drivers lic. & photo on it looked phony to me, his face didn't match his drivers lic. Valley spoke with a German accent but said he was born and raised in Los Angeles.

I advised Valley of his Miranda rights. He waived to talk to me by saying he'd tell me his story. He said he took a cab today from the airport in Los Angeles enroute to San Diego. He stopped at the Bank of Money in El Fuego to make a withdrawal from his savings account that he opened at another branch of the bank in San Diego last year. He came alone. He said he deposited two checks totaling approx. $7000 in his San Diego acct. 3 weeks ago. The checks were 2nd party checks from a friend of his in New York. Today he wanted to withdraw $2500 cash of that money.

After investigation it was ascertained that Valley (if that is his name) tried to steal $2500 from the Bank of Money. Three other males, probably friends of his split the location of the bank when I first arrived, & were subsequently apprehended. They all had German type accents and said they were from Los Angeles. I arrest the man purporting to be Clifford Valley on the warrants & additional charges relating to the attempt theft at the bank.

ADDITIONAL FACTS:

The Bank of Money is located at 10 Dollar Lane, El Fuego, CA 90001, (132) 555-1212. The person you talked to at the bank is Yolanda Sweet, the branch manager. According to Ms. Sweet, Valley and three other men came into the bank and got in the teller line. All had accents but she could not identify them but thought they were German or Russian. Valley presented his driver's license and wanted to withdraw $2500 from an account in San Diego. The account was in the name of C. Valley, but there was a hold on the account. You called the bank in San Diego and spoke to manager I. M. Honest who said that two stolen checks totaling $7000 had been deposited 4 days ago and had been returned to the bank. The warrant you found for Valley is F336514, for Forgery, Bail $150,000, issued on 5-2-98 in the West Court Judicial District.

4. Review your local newspaper to locate a criminal case in which an arrest was made and the case subsequently filed. Visit the court clerks office and see if the arrest report is available for review. If so, see if you can identify the probable cause for the defendant's arrest. Does the report follow the Rules of Narrative Writing?

## QUIZ

1. What is the most important thing to be established in an arrest report?
2. What are two basic types of arrest report formats?
3. What is the format for side headings for a straight narrative arrest report?
   a.
   b.
   c.
   d.
   e.
   f.
4. How should the narrative of an arrest report begin?
5. When should an arrest report be completed?
6. What is the minimum length of an arrest report?
7. Injuries to the suspect should only be explained if they are serious. True or False and explain.
8. When a suspect is arrested for multiple charges, only the most serious charge should be established with probable cause.
   a. True
   b. False
9. Generally speaking, the arresting officer should include an opinion in an arrest report. True or False and explain.
10. It is the arresting officer's duty to write a biased report that will insure a suspect is convicted. True or False and explain.

# 7 Issues in Writing

Several years ago, a fictional investigator became famous for his ability to question people and quickly get to the bottom of the matter at hand. Whenever the person he was questioning would stray off course with an answer, this no-nonsense investigator would tell them, "Just the facts," and the interviewee would soon after provide a key piece of information needed to complete the investigation. Although this is unlikely in real life there is a lot of truth to the expression *just the facts* as it relates to investigations.

## FACT OR OPINION?

For the most part investigations need to contain facts. Facts are the tangible things we use to make solid decisions, they are things that can be proven, and they are the nuts and bolts of a quality investigation. If facts are at one end of the spectrum, the other end contains things like suppositions, hunches, gut feelings, and opinions. There is a great deal of difference between a fact and an opinion.

A fact is something that can be proven. An opinion is a personal belief or judgment that cannot be proven. Although an opinion may be a belief or judgment shared by many, it is not something that can be used to prove certainty. A general rule of thumb is that investigative reports should contain facts only. Opinions should not generally be included, however, there are some occasions when opinions are necessary in an investigative report.

Many experienced investigators have the mistaken belief that the things they have learned through years of experience and their own common sense are facts that should be included in an investigative report. This occurs when an investigator describes how a suspect accomplished a crime even though there is no available evidence. An example is an auto burglary accomplished by the use of the car key to make entry. In this scenario the driver will park, hide the key somewhere on the car, and leave it unattended for several hours or days. When the owner returns and finds the car unlocked and the car stereo missing, he or she calls the police, and the officer who has been assigned to investigate this crime begins a search of the area. The investigator finds that the hidden key is still in place, there is no sign of forced entry, and no witnesses who can tell him or her how the suspect got into the car. Based on this the investigator decides that the suspect used a shimmying type of tool to unlock the door or some other type of lock pick and in his or her report writes:

The suspect gained entry by shimmying the door lock.

Based on the fact pattern the appropriate conclusion for the investigator to make is that the method of entry is unknown.

Another example of how an opinion is inappropriately used in an investigative report is the residential burglary. The suspect ransacks the house, turning drawers upside down and spilling their contents on the floor, emptying closets and throwing clothes onto the furniture and floor, leaving few things untouched. After the investigation, the investigator writes a report in which he or she reconstructs the crime and says with certainty which room the suspect entered first, which drawer was dumped out first, what items of property the suspect removed, and in what order. When questioned about how the investigator can be certain how the crime occurred, he or she will almost always say it is from experience. This may be a very desirable trait to have available, but it is not appropriate for a factual investigative report. In this example, it would be more appropriate to say that the house was ransacked, and leave it at that.

## The Expert Opinion

As mentioned earlier there are some instances when the inclusion of opinions is not only appropriate but necessary. The vast majority of these occur during arrests, when the expert opinion of the investigator establishes the corpus of a crime and therefore is the basis for an arrest. Two examples of this are arrests for being drunk and driving under the influence.

First, the common drunk is a situation that requires an investigator to understand and interpret the symptoms of alcohol intoxication and evaluate these symptoms against the applicable ordinance or law prohibiting such behavior. Once the investigator has made the determination that a person, Steve Drinker in this instance, is in violation of an applicable law, he or she establishes the corpus of such a crime by forming an opinion (based on all the available information) that Drinker was in violation and arrests him. The investigator might show this in his or her report by writing:

> Based on Drinker's objective symptoms of alcohol intoxication, I formed the opinion he was under the influence of alcohol to the point he could not care for himself or others and I arrested him.

The second situation in which an opinion is properly used to establish the corpus of the crime is in the case of driving under the influence. In this example a field investigation is done in which the investigator sees the symptoms of alcohol influence, and the effect of this alcohol influence on the person's ability to operate a motor vehicle in a safe and prudent manner. The investigation may include observations of improper driving, or behavior that is well outside the norm for persons who are not under the influence of alcohol, in like driving conditions. Based on this driving and the obvious signs of alcohol influence, the investigator forms the opinion that the driver, Steve Drinker, is operating a motor vehicle under the influence of alcohol and arrests him for that offense. This might be written:

> Based on Drinker's objective symptoms, his inability to operate a car, and the results of the field sobriety test, I formed the opinion that he was under the influence of an alcoholic beverage to the point that he could not operate a motor vehicle in a safe and prudent manner, and I arrested him.

## Documenting Responses to Miranda Rights

Another area where opinions show up time and time again is when investigators write about giving the Miranda admonishment and the resulting waiver. The timing and procedure to be used in giving someone their rights per the Miranda decision is more appropriately discussed in an investigation class, however, the matter of how to record this act is properly addressed here.

Regardless of how the matter is taken care of it is important for the investigator to be able to accurately recall how the event came to be even though it might be months later. One successful method is to read the Miranda admonishment to the person and quote their response to the two waiver questions. This could be written as:

> I read Drinker his rights per Miranda from a card I carry for that purpose. In response to the two waiver questions he said, "Yes" and "I'll talk to you," respectively.

By quoting the responses and referring them to the waiver questions you will remove the guess work from the process and will be able to testify with certainty as to what was said. The problem is that many investigators forget the importance of being able to accurately recall what the suspect said, and then fail to write it in their report. Instead of quoting the suspect, they write:

> I advised Drinker of his Miranda rights and he acknowledged and waived them.
> I gave Drinker his rights and he waived them in an intelligent manner.

Now, both of these examples have the classic ring of a good investigator to them, but they are disasters waiting to happen. In both cases the investigator failed to establish that the Miranda admonition was given properly or that the person who received the warning understood it and actually waived his or her right against self incrimination. What we have in this example is an investigator's opinion that the person heard the actual Miranda admonishment that they are entitled to, and then waived those rights. What would happen if the investigator was asked to repeat in court the exact words he or she used to give this admonishment, and in so doing, forgot or left out part of it? The end result might be the exclusion of any information the person told the investigator that helped establish a crime.

It would be better to be in a position of being able to read the rights from the very same card or form that was used to admonish the suspect and then be able to quote his or her exact words. This would allow the trier of fact to decide if the admonishment was given properly and if the waiver was an intelligent and voluntary one.

It is just as important to include a negative response to the Miranda admonishment in a report because it alerts all who read the report that the suspect does not want to talk about the charges against him or her. Just as in the waiver situation, it is important to use the exact words the suspect used when invoking his or her Miranda rights. One way this might be written is:

> I read Cleworth his rights per Miranda from a card I carry for that purpose. In response to the two waiver questions he said, "Yeah" and "I'm not saying anything until I talk to my lawyer," respectively.

## Documenting Field Show Ups

Still another area where the investigator's opinion seems to creep into reports with some regularity is during field identifications or show ups. These are times when a suspect is stopped near a crime scene and a witness is brought to the suspect to see if the person being detained is or is not the perpetrator. It is important to remember that the same reasoning applies to field identifications as to Miranda admonishments. It is the investigator's job to see that the words the witness used are accurately recorded in his or her notes and later used in the report to establish probable cause or, just as importantly, to eliminate someone from the suspect pool. When field identifications are done, witnesses should view the suspect separately, and the admonition and response should be available for later testimony. The words the witness used should be quoted when referencing what happened and not paraphrased or generalized by the investigator. The way this might be worded in a report is:

> I stopped one block from where Brown was being detained and read Fava the field admonishment from a card I carry for that purpose. I drove to within 20 feet of Brown and when Fava saw Brown he said, "That's him, there is no doubt in my mind."

By reporting the facts in this manner it allows the trier of fact to decide whether or not the witness made a good identification or not.

When people talk about good report writers, they are really talking about what constitutes a good report. A good report has all the right things in it and none of the wrong things. It meets the needs of those who read it, yet does not require a great deal of time to complete or read.

It is important to understand who reads reports and why. A long list of persons who do can quickly be developed. It includes supervisors, bureau commanders, division commanders, chiefs of police, district attorneys, defense attorneys, defendants, planners, city administrators, newspaper reporters, and the general public. Many of the reports an investigator prepares have the chance of ending up in the hands of just about anyone, and certainly the information in the report can end up under the eyes of anyone who can read.

## READER USE CONFLICT

The issue of who a report is being written for must be recognized because the people who review it may have different needs or desires. This is the issue of reader use conflict, and understanding it can make the task of completing a quality report much easier.

The investigator should remember this because it can help him or her understand what needs to be done to achieve a quality investigation. An important step is to place yourself in the position of those who might be the biggest critics of your work and look at the entire situation through their eyes. Look closely at what is developing and if a potential problem appears, take care of it by doing what is appropriate given the legal constraints, rules, and procedures you are operating under. By taking care of potential problems at that stage of the investigation, completing the report later is no problem because all you have to do is write about what happened. In Most criminal investigations there are several common issues, some of which are:

### Why Are You Investigating This in the First Place?

If you recall the rule on how to begin the report this is an easy issue to take care of. By starting the narrative with the date, time, and how you got involved, the answer is right there in the first sentence of the report.

### How Did You Focus on the Suspect?

Writing in chronological order allows you to lay out step by step what happened and the things you did that ultimately led to the suspect. It may be the result of good investigative work, or it may be the result of good luck. Regardless, the narrative should contain nothing but the truth, and the truth should be written in factual terms.

### Why Did You Detain the Suspect?

Probable cause is the answer here as well. Can you write what the facts were that led you to believe it was necessary to conduct a search? If so you have passed the test and eliminated this issue.

### Did You Establish a Corpus?

Do you know what the corpus is for the situation? If not, can you find out by reviewing the appropriate criminal codes or jury instructions? You need to know what constitutes a crime and how to establish it in writing.

### What Did You Do with the Evidence?

Did you establish a good chain of custody? Are you able to account for the evidence for every second since it was seized? Have you handled it properly? Is it marked and packaged in an acceptable manner?

### Why Did You Arrest the Suspect?

This, once again, is in the area of probable cause. It is not enough to have had a good reason, you must be able to write about it so that whoever reads the report will be able to also understand your reasoning.

### Was Miranda Documented in the Proper Manner?

Will you be able to testify about what was said and who said it when the case reaches court? Is what you wrote free of opinions? If not you need to rethink how to make it factual in future cases.

By placing yourself in the role of the devil's advocate, you can see where improvement is needed and work toward eliminating the problems that might arise. If you are able to recognize a problem at the investigative stage and arrive at a successful way of dealing with it, the task of writing the report is nothing more than putting down on paper what you did. If you are able to eliminate the issues of

**Figure 7-1**

report writing at the time they come up, you will be establishing good investigative habits. This will allow you to do a thorough job throughout your career and achieve the goal of investigating, which is to find the truth.

## REVIEW

1. Facts are things that can be proven.
2. Opinions are personal beliefs that have limited use in investigative reports.
3. Probable cause can be established by expert opinions.
4. An appropriate method of giving the Miranda admonishment is to read the admonishment and quote the person's response.
5. Issues are potential problems.
6. Different readers have different needs when it comes to what they want and expect in a report. This is known as *"reader use conflict."*

## EXERCISES

1. Read the following report and identify any issues and discuss in class. Rewrite the report following the Rules of Narrative Writing. For the purposes of the report, the paramedics' names are R. Ho and G. Najjar. Captain Ho is in charge of the crew and the one you spoke to.

Type:   Death Report      Date:   2-22-95
Location:   17221 Main Street, El Fuego
Date and Time of Occurrence:   2-22-95 1500 Hours
Victim:   Brown, Roger Anthony Male, white, 49 years
Address:   17221 Main Street, El Fuego
How Occurred:   Liver Failure
Discovered or Witnessed By:   Brown, Marian NMN
Address and Phone:   17221 Main St., El Fuego (123) 555-1212
Did Ambulance Respond:   No
Doctor on Duty:   No
Victim Moved To:   The Neptune Society
Moved By Whom:   The Neptune Society
Relative of Victim:   Marian Brown, Wife
Relative Notified:   Yes
By Whom:   The R/P Marian Brown
Victim Under Care of Private Physician:   Yes, 4 years
When Private Doctor Last Seen:   1-29-95
Name of Private Doctor:   Ben Togani, MD, El Fuego Med. Center
Nature of Ailment:   Cirrhosis of the liver
Was Doctor Present:   No
Victim Pronounced Dead By:   El Fuego Fire Department

Death Certificate to Be Signed By:   Coroners Office
Coroner Notified By:    EFPD
Describe Any Drugs Found:    Lasix 80 mg., Cpirenolactone
                              25 mg., Propranolo 10 mg.

DETAILS:

At approximately 1510 hours this date, I was dispatched to the Main location. Upon my arrival there, I contacted the R/P, wife of the deceased, Marian Brown.

She stated that at approximately 1440 hours, this date, she checked on her husband who was in the upstairs master bedroom and at that time, although he was feeling ill, he was resting in bed, dressed in a pair of white shorts. She subsequently left the location to go to the store. When she returned home at approximately 1500 hours, she went upstairs to check on her husband; at that time she discovered him lying in a supine position, deceased. She immediately contacted El Fuego Paramedics. The El Fuego Paramedic team responded to the location after being dispatched at 1501 hours. Upon their arrival there, a fire unit pronounced the deceased dead and notified El Fuego Police Department at 1510 hours.

Sergeant Smith responded to the Main location and, after being advised by the reporting officer of the discovery and the prior physical condition of the deceased, Sergeant Smith subsequently left the location, leaving the reporting officer in charge.

I then recontacted the R/P, Mrs. Brown. She stated to me that her husband has been suffering from severe cirrhosis of the liver for approximately four years. He has been treated by numerous physicians at the Renal Clinic portion of the El Fuego Medical Center in the city of El Fuego. The last time he was at the facility, which was on January 29th, he saw Doctor Togani who had prescribed several of the above listed medications. Mrs. Brown further indicated that on or about the 15th of January, Mr. Brown had been admitted to the El Fuego Medical Center hospital where he spent two days due to severe blood and potassium level depletion. She stated upon his return to his residence, he did not follow the advice of his physicians and continued drinking alcoholic beverages on a daily basis.

I then responded to the upstairs master bedroom located upstairs in the northeast corner of the residence. There, I entered the master bedroom and observed the deceased to be lying in a supine position with his right shoulder and back against the mattress, along with the right side of his head lying against the mattress. I could observe blood emanating from the deceased nose and mouth, a small pool of which had formed on the mattress. Upon closer examination of the body, I observed severe swelling in the lower abdomen and upper stomach area of the deceased. There was no evidence of fowl play on the body and the body was still clad in a pair of white shorts.

Crime Scene Investigator James #1207, arrived at approximately 1555 hours. He took photographs of the deceased and the surrounding bedroom area.

Upon close examination of the bedroom, hallway and bathroom, it was apparent to the reporting officer that after the R/P, Mrs. Brown, had left the location, the deceased had used the upstairs bathroom which is located next to the master bedroom. A small amount of blood was observed on the cabinets and on the floor of the bathroom. From there was a small spotting of the carpeting leading to the area of the bed where the deceased, after passing away, fell backwards, lying in a supine position, causing a small amount of blood to form on the bed.

Grande County Coroner's Deputy Rolly P. Wedges responded to the location at approximately 1625 hours. At that time, after close examination of the deceased and the above described evidence, he substantiated the reporting officer's conclusions. For details regarding Deputy Coroner Wedges report, see Coroner's case number DB-95-1631.

Deputy Coroner Wedges took charge of the body and indicated that there would be no autopsy and the mortuary of the Neptune Society had been notified and would respond to take charge of the body. At that time Deputy Wedges indicated that he would remain with the body and take charge of same until their arrival.

2. Read the following report and identify any issues and narrative writing problems. Rewrite the report following the Rules of Narrative Writing and address the issues previously identified. For the purposes of the exercise, assume that Linton is driving under the influence of alcohol and that it is lawful to arrest him, and that once he has been arrested he must provide a blood, breath, or urine sample to determine his level of intoxication.

| | |
|---|---|
| ARRESTED: | Linton, Dwight K. DOB 5-14-63 |
| CHARGES: | Driving under the influence |
| LOCATION: | Parkside and Stark, El Fuego |
| DATE & TIME: | 3-7-96 2245 Hours |
| OFFICERS: | M. Colin #4787 and R. Sproul #12298 |
| VEHICLE: | 1989 Porsche 924, blue, license QUIK ONE |
| WITNESS 1: | Books, Linda L. DOB 6-21-65 |
| | 2555 Stark #G, El Fuego |
| WITNESS 2: | Budding, Frank D. DOB 8-13-54 |
| | 2565 Stark #K, El Fuego |

DETAILS:

At approximately 2239 hours, I received a radio call of a disturbance at 2555 Stark, Apartment G. The call stated that the R/P's ex-husband just hit the R/P and the subject was now taking the R/P's blue 1989 Porsche 924. R/P wants to prosecute the subject who is taking her car, one Dwight K. Linton.

After receiving the above call, I arrived at the location at approximately 2245 hours. The apartment building at 2555 Stark is on the southeast corner of Stark and Parkside. As I pulled up I parked on the northeast corner of Parkside and Stark. At that time I was approached by witness Budding and a few other people who were standing on the corner. Mr. Budding and the other people told me that they had seen a subject leave at a high rate of speed from the area and offered to be witnesses if any were needed. They advised me that something had happened directly across the street (pointing to apartment G at 2555 Stark). They told me they did not know what had gone on. I advised them that the lady at that location had called about a problem with her ex-husband or boyfriend.

I then got out of my car and was getting a clipboard for a possible stolen car report, when the suspect drove the above listed blue Porsche southbound on Parkside. As he was doing so he was also honking the horn. Mr. Budding and several other subjects then pointed the car out to me stating, "There he is." I still had not contacted the R/P prior to this time. The only thing I knew on the call was that this was apparently the ex-husband of the R/P. Seeing that the car was now evidently being returned, I walked over to contact the ex-husband (arrestee). The suspect was now parking the vehicle on Parkside, just south of Stark, on the east side. He parked his vehicle parallel at the curb just north of the driveway which goes behind the apartment complex at 2555 Stark.

As I approached the vehicle I observed him get out of the vehicle and then close the driver door. I observed his gait to be extremely unsteady and he was staggering and holding onto the car for balance. As I walked up to him and stood next to him, it was apparent that he did not even notice my presence (I'm standing 2 to 3 feet from him). I watched him for approximately a minute and a half attempt to put a key into the door of the vehicle and lock it. He then looked up at me and I noticed that his eyes were bloodshot and watery and his face was flushed. He had a very strong odor of an alcoholic beverage about his breath and clothing. Not knowing the R/P's name, I then asked him if he was the ex-husband that I had been called on. He then mumbled at me words to the effect that he probably was, and I noted that his speech was extremely thick and slurred. It was very apparent to me that the suspect was drunk and was unable to care for his safety. I had also observed him just drive up in a vehicle in this condition. I then asked for his driver's license. He looked at me blankly and began to stagger towards the sidewalk. I let him walk out of the street and then took hold of his arm on the sidewalk, stopping his forward motion. I again asked him for his driver's license. He then looked at me and told me that he didn't have one. I told him that I was arresting him for driving under the influence of alcohol, at which time he mumbled something incoherently at me and began to try to walk away again. I again grabbed his arm and he began trying to twist away from me.

The suspect's balance was extremely uncoordinated and it made him very hard to handle physically because of his size (approximately 6 feet tall, 230 pounds). I then was able to put his right arm in a wrist lock behind his back and sit him down to the ground and then roll him over onto his stomach. I then placed his other hand behind his back and handcuffed him. I then escorted him (holding him up as he staggered down the sidewalk) to my police car. He was complaining to me as I walked him down the sidewalk that he was disabled and that I had hurt him. He also, as I reached the police car, asked me what I had arrested him for. I again told him that he had been arrested for driving under the influence. He asked me why, and I told him I had seen him drive up in his car. He then stated to me, "What car?" At that time I realized that his comment drew somewhat of a chuckle from a small group of people who were standing on the corner next to my police car; Mr. Budding was one of the citizens in that group. Officer Sproul #12298, then arrived on the scene and stood by, with the suspect in my vehicle, while I contacted the R/P. The R/P (witness Books) at that time did not know that I had arrested her ex-husband and that her car had been returned.

She told me that approximately 10 minutes ago her husband had taken the car without her permission and had physically assaulted her in doing so. She told me that the police had been called to her house twice before earlier this evening regarding her ex-husband who was drunk. She advised that both times the police responded her husband had left the area and had not been found.

She told me she had first came home at approximately 6:30 this evening and the suspect at that time was extremely drunk. I asked her if he was drinking after she arrived home, and she had not been home most of that time, but had been in and out. She did state that he had taken a gallon of rum with him and he had not returned with it prior to taking her car and she assumed that he had been drinking it.

She advised me that her husband, just prior to taking the car, came in to the house (apartment G) and had wanted her keys. She did not want him to take her car, at which time the suspect forcibly took the keys and went out to the parking area behind the apartments. She advised that the suspect got in her vehicle and she had attempted to pull the keys away from him. The suspect had backed up, with witness Books leaning in the driver's door, attempting to get the keys out. The suspect, after backing up a few feet, had pulled forward and to the left to negotiate driving out of the driveway westbound onto Parkside. The driveway is approximately 10 to 12 feet wide and has cinder block abutments on both sides. She advised that he did not negotiate his exit properly and the left door of the car struck one of the cinder block abutments and knocked it down. She stated at that time she was able to grab the keys from him and run back around the front of the apartment. She attempted to get into the apartment of a friend of hers (apartment C, downstairs from her apartment). She stated as her friend answered the door, suspect Linton came up and grabbed the keys back from her and went back around to the car. She was unable to go after him this time, but heard him leaving the location on Parkside (spinning the tires and screeching). She then called the police and I was dispatched.

I checked the victim's vehicle with her. I noted there was a small amount of scraping damage on the driver's side door. She also pointed out some more very obvious scrapes on the right side door and the right rear quarter panel of her car, which she stated the suspect had also done since he had just taken the car. It was unknown what he had scraped and no paint transfer could be seen. She advised it was very possible that after he had re-taken the keys from her and had successfully left in the car, he may have very well struck one of the other cinder block abutments because of the odd position that he had came to a stop in prior to her getting the keys. At this time she did not want to pursue the matter of assault or auto theft. She did advise me that the suspect did have some sort of back disability from some prior condition.

During my transportation the suspect constantly told me that he was going to sue me and my family for false arrest. He kept telling me that I had not used my red lights and therefore I could not arrest him for a "502." He told me that I could only arrest him for walking drunk because he had already parked the car and was on the sidewalk when I arrested him.

Prior to getting to the station, he also demanded a blood test. Due to his demands for a blood test, I did not initially advise him of his obligation under the implied consent law, but advised him that if he did wish to take a blood test, he could. I did ask him briefly if he wished to take a breath test or a urine test instead, and he demanded that he wanted a blood test only.

Just prior to 2350 hours, I escorted the suspect into the El Fuego Jail infirmary to have a blood test taken. In the room was Jail Nurse Brown and Jailer Roland. The suspect at that time told me he wanted a cigarette or he would not submit to a blood test. I told him that there was no smoking in our jail. He then told me that he wanted a lawyer before taking any blood test and I advised him that it was not his right. He then began pounding on the table and it appeared as if he was going to bolt out of his chair and attack. Myself and Jailer Roland then restrained him back into his chair and told him to calm down. I told him I would read him his rights under the implied consent law if he wished to hear them. I then read verbatim from state form 13353 to the suspect. He then seemed satisfied that he had been told his rights regarding tests and then said that he now wanted to take a blood test. He then submitted to a blood test without any further incident.

At 2350 hours, Nurse Brown took two vials of blood from the suspects right inner arm. Zephiran was used as the sterilizing agent. Nurse Brown retained the blood vials for processing. Brown marked, packaged, tagged, and booked the evidence.

During the blood test, the suspect kept reiterating at me that I had not used my red lights to stop him and therefore I had no case. He then again stated that he had already parked the car and was on the sidewalk before I had arrested him and therefore I could only arrest him for drunk. He also stated words to the effect that his blood test would prove this out. After the blood test, the booking was completed without any further incidence.

End of report.

M. Colin #4787

3.  Assume you are a veteran officer and your partner, a rookie named Roger Cleworth, comes to you for advice. It seems that an arrest report he wrote was "KICKED BACK" by the sergeant because there is a Miranda problem. You read the report and find that Cleworth did a nice job of establishing probable cause and wrote a good report with one exception, the Miranda admonishment. Cleworth wrote:

"I advised Rand of his Miranda rights which he understood and waived."

When you talk to Cleworth, he tells you that he really read the Miranda admonishment to Rand and can tell you the exact words Rand used when he waived his rights. Cleworth looks at his notebook and shows you that Rand said "I do" when asked if he understood his rights and "Yes" when asked if he would talk. What advice would you give your partner and how should he write this piece of his report?

## QUIZ

1.  What is a fact?
2.  What is an opinion?
3.  Opinions should not be used in police reports.
    a.  True
    b.  False

4. What is an issue?
5. Give an example of an issue.
6. When should an opinion be used in a police report?
7. What is reader use conflict?
8. What are two incidents in which opinions might be appropriate in a police report?
9. The chapter described eight common issues in police reports. Name four:
   a.
   b.
   c.
   d.
10. One way of minimizing issues in a report is to identify them early in the case and work to correct them during the investigation. Through whose eyes should you look at your case to identify these issues?

# 8  Writing Warrants

At some point in your investigative career you will encounter a situation where you need some information that is not readily available. Your need might be business records or access to a crime scene that requires more than consent—in short, you will need a search warrant. When this situation arises, you will need to have a working knowledge of search warrants and the process needed to obtain one. The purpose of this chapter is not to provide an all-encompassing report on this investigative tool but rather to establish a writing strategy for completing the necessary documents to obtain, serve, and return a search warrant.

## SEARCH WARRANTS

One of the assets a good investigator must have is a working knowledge of search warrants. In today's world, a lot of valuable information can be obtained, but it is getting more and more likely that the person or business in possession of what you need will not volun-

**Figure 8-1**

tarily part with it. This makes it essential that you know what to do and how to do it, if and when you are faced with this challenge. This section is not an attempt to cover every aspect of search and seizure as it relates to search warrants, nor to qualify you as an expert in the area of search warrants. Rather it is to give you a basic overview of what a search warrant is, how to write one, and what the process is to obtain, serve, and return one.

What is a search warrant? Although exact definitions may vary from jurisdiction to jurisdiction, the *thirty word version* is that a search warrant is an order from a magistrate to a peace officer, to go to a particular place, search for a particular thing, and if found, bring it back and show it to the magistrate who issued the warrant. A search warrant can be issued only when probable cause exists, and your job as an investigative report writer is to put the probable cause into the proper format. Generally the probable cause must show that one of the following things has recently happened or is occurring:

1. When Property is stolen or embezzled.
2. When the property or things were used as the means of committing a felony.
3. When the property or things are in the possession of any person with the intent to use it as a means of committing a public offense, or in the possession of another to whom he or she may have delivered it for the purpose of concealing it or preventing its being discovered.
4. When the property or things to be seized consist of any item or constitutes any evidence that tends to show a felony has been committed, or tends to show that a particular person has committed a felony.
5. When the property or things to be seized consist of evidence that tends to show that sexual exploitation of a child has occurred or is occurring.
6. When there is a warrant to arrest a person.

A search warrant consists of three distinct parts. They are the search warrant, the affidavit in support of the search warrant or statement of probable cause, and the return to the search warrant.

## THE WARRANT PROCESS

When the need for a search warrant exists, the affiant, who is the person that prepares the warrant, compiles all of the facts and evidence into written form, properly completes the search warrant forms, and presents the search warrant and affidavit to a deputy district attorney for review. Once the deputy district attorney agrees that the information in the warrant and affidavit support its issuance, the affiant takes the warrant to a magistrate. The magistrate

STATE OF CALIFORNIA - COUNTY OF ORANGE    SW NO. _____

# SEARCH WARRANT AND AFFIDAVIT
# (AFFIDAVIT)

_____ declares under penalty of perjury that the facts expressed by him/her in
(Name of Affiant)
the attached and incorporated **Statement of Probable Cause** are true and that based thereon he/she has probable cause
to believe and does believe that the articles, property,and persons described below are lawfully seizable pursuant to Penal
Code Section 1524, as indicated below, and are now located at the locations set forth below.  Wherefore, affiant requests
that this Search Warrant be issued.

_____ ,  **NIGHT SEARCH REQUESTED:    YES [  ]    NO [  ]**
(Signature of Affiant)

# (SEARCH WARRANT)

THE PEOPLE OF THE STATE OF CALIFORNIA TO ANY SHERIFF, POLICEMAN OR PEACE OFFICER IN THE COUNTY OF ORANGE:
proof by affidavit, under penalty of perjury, having been made before me by_____
–
(Name of Affiant)
that there is probable cause to believe that the property or person described herein may be found at the locations set forth
herein and that it is lawfully seizable pursuant to Penal Code Section 1524 as indicated below by "x"(s) in that it:
_____ was stolen or embezzled.
_____ was used as the means of committing a felony.
_____ is possessed by a person with the intent to use it as means of committing a public offense or is possessed by
       another to whom he or she may have delivered it for the purpose of concealing it or preventing its discovery.
_____ tends to show that a felony has been committed or that a particular person has committed a felony.
_____ tends to show that sexual exploitation of a child, in violation of Penal Code Section 311.3, or possession of matter
       depicting sexual conduct of a person under the age of 18 years, in violation of Section 311.11, has occurred or is
       occurring.
_____ there is a warrant to arrest the person.

YOU ARE THEREFORE COMMANDED TO SEARCH:    (Premises, vehicles, persons)

FOR THE FOLLOWING PROPERTY OR PERSONS:

AND TO SEIZE IT/ THEM IF FOUND and bring it/ them forthwith before me, or this court, at the courthouse of this court.  This **Search
Warrant** and **Affidavit** and attached and incorporated **Statement of Probable Cause** were sworn to as true under penalty of perjury
and subscribed before me on (date) _____ , at _____ A.M. / P.M.  Wherefore, I find probable
cause for the issuance of this Search Warrant and do issue it.

KNOCK-NOTICE EXCUSED:☐

**Figure 8-2**   (As described in California Penal Code)

swears the affiant in, much as a witness is sworn in before testifying, and reviews the warrant and affidavit. If the magistrate agrees that the warrant and affidavit have established probable cause for a search, the affiant signs the document, and then the magistrate signs. At this point the warrant must be served and returned to the court within 10 days.

After serving the warrant, and seizing the property listed in the warrant, the affiant prepares the return to the warrant. The return is a written list of all evidence seized pursuant to the service of the warrant. Once the return is prepared, the affiant takes the original search warrant, the affidavit, and the return back to the magistrate and shows the magistrate what was seized. Both the affiant and magistrate sign the return and the affiant then gives all three parts of the warrant to the court clerk, who files it. Generally speaking, at this point all the information contained in the search warrant, affidavit, and return become public record, and anyone may obtain a copy of them.

## HOW TO WRITE A WARRANT

The search warrant is the part of the document that contains the description of the place to be searched, and a description of the evidence you are looking for.

The description of the premises to be searched must be so complete that any peace officer could pick up the warrant, read the address and be able to find the location to be searched. This can be accomplished by describing the premises with the address, city, county, and state, followed by a physical description of the property including style of construction, color, where the address numbers are attached to the building and the geographical location of the structure in relation to fixed reference points. For example:

> 123 Elm, City of Long Wave, County of Los Angeles, State of California, a single story wood and stucco structure with a brown shake roof. The front door of the residence faces south and the numbers 123 are attached to the fascia directly over the front door. There is an asphalt driveway along the east side of the residence, and the house is the fourth structure west of Vine Avenue on the north side of the street.

If the premises or location to be searched is an open field, the description might be:

> An undeveloped lot on the southwest corner of Beach Boulevard and Main Street, City of Long Beach, County of San Diego, State of California. The lot is bordered on the west by a six-foot tall block wall, on the north by the south curb line of Main Street, on the east by the west curb line of Beach Boulevard, and on the south by the north wall of the Acme Toy Company.

The entire lot is visible from Main Street and Beach Boulevard and has no structures or trees on it. Although there is no known address for the lot it is in the 8000 block of Main Street and the 200 block of Beach Boulevard.

In some cases, searches must take place in remote areas, far from developed streets and highways. In these cases your description might include a picture or a reference to a map to help find the exact location. For example:

An abandoned wooden shack one half mile north of Fire Road 6 and two miles west of Highway 14 in the Southern National Forest, county of Los Angeles, State of California. The area is accessible by air and four-wheel drive vehicles and appears to be without utility connections. The area of the premises to be searched is shown on page 1134 of the 1997 edition of the Maps-R-Us Road Atlas at coordinates KT-89. A copy of this page has been marked with an X to show the location of the shack and is attached herein and incorporated as EXHIBIT A. A photograph of the shack has been attached herein and incorporated as EXHIBIT B. The shack is a single story, wooden structure with no windows and a door on the west side.

If your investigation requires that you search a vehicle your description might be:

A 1946 Chevrolet Panel Truck, dark gray in color with California license plate 1G63410. The vehicle is missing the left front fender and the right rear wheel. It is parked on Elm Street, south of Orange, in the City of Chico, County of Butte, State of California. The car is sitting on concrete blocks with the wheels off the ground. The vehicle was last registered in 1964 and there is no current registered owner information available per the Department of Motor Vehicles.

Obtaining business records such as telephone subscriber information or bank account information is likely to be part of any long-running investigation and should be looked at as any request for information. One difference in serving a search warrant for business records is that you are usually not going to actually conduct a search of the business. Normally, you will notify the custodian of records ahead of time and let the custodian know what records you are looking for. Your search warrant is then presented to the proper authority and the records are turned over to you. If you need to access a safe deposit box at a local bank, your description of the premises to be searched might be:

Safe deposit box 1225 located in the vault of the Bank of Money, 121 Dollar Drive, City of Cash, County of Coin, State of California. The Bank of Money is in the Willow Branch

Shopping Center and sits on the northeast corner of Dollar Drive and Walnut. The Bank of Money is between the Good Burger Restaurant and the Lifetime Appliance store. The bank is a three-story masonry building, brown in color with the numbers 121 attached to the east side of the building directly over the door.

If you are looking for subscriber information at a phone company, perhaps the description would look like:

West Coast Telephone Company, 6426 Cellular Drive, City of Big Tree, County of Oakland, State of California. The requested records are kept by the custodian of records in Room 100 on the first floor. The building is on the south side of Cellular Drive between 64th and 65th streets. I do not intend to conduct a search of the premises, but rather to turn over the search warrant to the custodian of records and receive the requested information.

Remember, the test of whether your description of the premises to be searched is good enough is whether any police officer could read your description and then go find the location or premises to be searched.

## DESCRIBING WHAT YOU ARE LOOKING FOR

The description of the property you are looking for must be good enough that the average person could read the description and pick the item out of a group of similar objects. There are thousands of items that can be the subject of a search warrant and this book is not designed to provide examples of them. The rule of thumb here is that the property you are looking for must be described with reasonable particularity, and the need for the items you are seeking must be supported by probable cause in the affidavit. Local prosecutors may have preferred formats or descriptions of the more common items sought such as phone records, bank account information, narcotics, and bookmaking records. It is always advisable to seek their input to insure your warrant is complete and sufficient.

### Writing the Affidavit

The affidavit, or statement of probable cause, is the part of the document that contains all of the facts, information, and evidence that establishes the probable cause for the issuance of the warrant. In addition to the facts of the case, the affidavit also includes a description of the affiant as a way of introduction to the magistrate.

The first part of the affidavit should be your introduction. Although there is no ironclad formula for the opening, it should

_____,     NIGHT SEARCH APPROVED:☐
(Signature of Magistrate)

Judge of the **Superior / Municipal** Court, _____ Judicial District, Dept. / Div. ____

F026-    (Rev.1/98) Pg.1                                                                f:\forms\sw-af1.198

STATE OF CALIFORNIA - COUNTY OF ORANGE     SW NO. _____
ATTACHED AND INCORPORATED

# STATEMENT OF PROBABLE CAUSE

Affiant declares under penalty of perjury that the following facts are true and that there is probable cause to believe, and affiant does believe, that the designated articles, property, and persons are now in the described locations, including all rooms, buildings, and structures used in connection with the premises and buildings adjoining them, the vehicles and the persons:

**Figure 8-3**    (As described in California Penal Code)

include your job and title, how long you have been in the assignment, prior experience, education, and training. You should also include practical experience in the particular type of case that is being investigated and if and how many times you have testified as an expert in court regarding similar types of cases. All of this helps establish your expertise in the eyes of the judge who is reading your warrant. Beginning the first paragraph with your name is not necessary because it is understood that the person making the application for the warrant is the affiant. Some examples of these introductions include:

> I am a member of the Big City Police Department and have been so employed for the past six years. For the past two years I have held the position of Detective and have been assigned to the Burglary Unit. I am a graduate of the Big City Police Academy where I received instruction in investigating the crime of burglary. I hold an Associate of Arts Degree in Administration of Justice from Long Beach City College and have attended a 40-hour Basic Investigation class sponsored by the California Department of Justice and have investigated more than 200 burglaries. I have arrested more than 20 suspects for burglary and receiving stolen property and have spoken with them about their crimes and criminal habits. Based on my experience and training I am familiar with the types of burglaries being committed in this area and the common motives of burglars.

An example of someone who is assigned to a narcotics unit might have an introduction something like this:

> I have been employed by the Oak Tree Police Department for five years and have been assigned to the Special Investigations Unit for the past three years. During the first three months of this assignment I attended an 80-hour Narcotic Investigation class sponsored by the Drug Enforcement Administration, and a 40-hour Narcotic Recognition class sponsored by the Federal Bureau of Investigation. I have had contact with more than 200 narcotics users in the Oak County area and discussed with them the use and sales of heroin, cocaine, and other controlled substances. I have had personal contact with more than 50 persons who use and sell heroin in the Oak County area. I have received field training by senior officers in the field of controlled substances from Lieutenant D. Roman, Porter Police Department; Sergeant J. Pounds, Central Police Department; Officer G. Najjar, Winston Police Department; and Officer R. Ho of the Pico Police Department. I have participated in more than 150 narcotic investigations and have seized cocaine on more than 200 occasions and have testified as an expert in the field of narcotics in the Municipal Court of Oak County on 13 occasions. Because of my experience and training, I am familiar

with the trafficking of narcotics in the southwest portion of the United States and the methods used by users to acquire drugs and the methods used by sellers of controlled substances to distribute them.

Someone working in a vice unit might have an introduction like:

I am a police officer in and for the city of Long Branch and have been so employed for the past 13 years. I currently hold the rank of Sergeant, and I am the supervisor in the Special Crimes Unit with the responsibility of supervising all vice and special assignment investigations. During my career I have worked as a patrol officer, narcotics investigator, patrol sergeant, and watch commander. I have worked as an investigator for a total of six years and have investigated over 2000 cases. I have worked in undercover and administrative assignments, purchased illegal drugs and participated in hundreds of arrests for violations of narcotic and vice offenses. I hold a Bachelor of Science in Criminology from the University of California, and I am a graduate of the Oak County Peace Officers Academy. I have attended more than 30 specialized training classes and schools sponsored by the California Department of Justice, California Narcotics Officers Association, the Federal Bureau of Investigation, and the Drug Enforcement Administration. I have worked hundreds of cases with other experienced investigators, and I am familiar with the law as it relates to conspiracy, narcotics, and vice offenses. I have testified as an expert in the Municipal and Superior Courts of Oak County on at least 10 occasions.

While it is helpful to have a strong academic and professional background including recognized certifications, this is not necessary to be an affiant and prepare a search warrant. Everyone has to start somewhere and sometime, and every affiant's introduction will be different. What is important is to have a logical, understandable, and well written introduction for your warrant. The first step in preparing your introduction is to put together a training record of all the classes and training sessions you have participated in, along with copies of the certificates you received. This is vital as you may be questioned in court as to the validity of your claims and having the proof in hand is necessary. It also gives you a way to review the extent of your training and attend classes where necessary to strengthen any areas that are lacking. Once you have assembled the training record, start writing. It may take a couple of attempts to get an introduction that sounds good and accurately reflects your personal situation. The rule of thumb here is not to exaggerate or overstate your qualifications. Understate your experience rather than go overboard.

Once you have introduced yourself to the magistrate, you complete the affidavit by presenting the facts, information, and evidence about your investigation. This can be accomplished by using the Rules of Narrative Writing and starting this part of the affidavit with the date, time, and how you got involved in the investigation. The rest of the affidavit will explain what happened and how you gathered the evidence you have. Your story must establish the probable cause for believing the facts are true.

If you have occasion to incorporate other reports, documents, or photographs into your affidavit, you may do so by introducing the matter in your story and then incorporating it into your warrant by reference. For example:

> During my investigation I learned that a similar crime occurred in the adjacent city of Rosarito and that suspects matching the description of the suspects in this case were involved in that crime. A report by Rosarito Police Department Detective John Brown, consisting of 11 pages and documented under case number 97-30481, is attached herein and incorporated as EXHIBIT A.

By referencing this report, the information in it becomes part of your affidavit. All that remains to be done for this type of inclusion is to mark each page of the attached Exhibit with the appropriate label. For example:

> EXHIBIT A Page 1
> EXHIBIT A Page 2

If you include the opinion of another expert in your affidavit to help establish the probable cause, you will need to introduce this person to the magistrate just as you did for yourself. Such a situation might look like this:

> I reviewed the facts of this case with Billings Police Department Investigator R. D. Moran. Investigator Moran is a Police Officer for the city of Billings and has been so employed for the past 23 years . . . .

Once you finish establishing your expert's qualifications, you will want to include the person's opinion as to why they think the facts of your case establish probable cause.

When you have completed the story of your investigation, the affidavit is almost complete. All that remains is for you, the affiant, to draw some sort of conclusion that the property you are looking for is inside the premises you want to search. This is usually done by saying so in a short paragraph. The wording of the closing paragraph will vary depending on the type of crime and how solid your facts are. There is no minimum or maximum length to the affidavit. The test here is this: Does the information in the affidavit establish probable cause to search? If it does, the warrant is ready to present to a magistrate for review.

## THE RETURN TO THE WARRANT

The Return to the Warrant consists of a fill-in-the-blanks type of face sheet and a listing of the property and evidence seized during the search. A technique used by many investigators is to list the evidence and property on their agency's property report and attach it to the return. It might be referenced in the return as:

Refer to the attached property report, case number 97-63142, which is attached as EXHIBIT A.

This written inventory of the seized property is taken back to the magistrate who issued the warrant along with the original warrant and affidavit.

If your agency does not have a format for this purpose, consider using the one described in chapter 4, and list the evidence seized by category. As discussed, the categories could be:

GUNS
MONEY
DRUGS
ITEMS WITH SERIAL NUMBERS
ITEMS WITHOUT IDENTIFYING NUMBERS

Within each of these categories, list all of the items that belong in it. Let's say you served a search warrant and found two guns, $3684 in cash, a stolen television, and a kilogram of marijuana in a large plastic bag. Your evidence report, including a chain of custody might look like this:

1. (1) Smith & Wesson model 19, .357 caliber revolver, blue finish with black rubber grips. Serial number G61141. Found under bed in northwest bedroom by Smith.
2. (1) Walther PPK, .380 caliber, brown finish with brown plastic grips. Serial number unreadable. Found under the west end of couch in living room by Wilson.
3. $3684 in U. S. Currency consisting of (30) Hundred, (12) Fifty, (8) ten, and (4) one dollar bills. Found in toilet water tank by Searle.
4. (1) clear plastic bag, tied at the top with gray duct tape containing marijuana. 1026 grams gross weight. Found inside the kitchen sink cabinet by Najjar.
5. (1) Sony 27 inch color television, black case, serial number TZ641761193G. Found in the garage along the west wall by Ho.

I collected all of the items from the finders and kept them until I marked, packaged, tagged and booked them into the Big Tree Police Department evidence room.

## STATE OF CALIFORNIA - COUNTY OF ORANGE
# RETURN TO SEARCH WARRANT

_____, being sworn, says that he/she conducted a search pursuant to
(Name of Affiant)
the below described search warrant:

Issuing Magistrate: _____,

Magistrate's Court: **Superior/Municipal** Court, _____, Judicial District

Date of Issuance: _____,

Date of Service: _____,

and searched the following location(s), vehicle(s), and person(s):

R-1

**Figure 8-4**   (As described in California Penal Code)

**Items attached and incorporated by Reference:   YES [   ]   NO   [   ]**
I certify (declare) under penalty of perjury that the foregoing is true and correct.

Executed at _____ , California   _____
____

Date:_____ at _____ [A.M.]   [P.M.]             Signature of Affiant

Reviewed by : _____   Date: _____ at _____ [A.M.]
[P.M.]
            (Signature of Deputy District Attorney)

F026-    (Rev.7/97) Pg.3                                        Page ___ of ___

**Figure 8-5**   (As described in California Penal Code)

The main thing to remember is that the description of the evidence and property must be so good that the average person could read your description and pick the item out from several similar items. You must also be sure to *establish a chain of custody* in the property or evidence report.

The preparation, service, and return of search warrants is not something you are likely to do everyday, but you should be prepared to do so at any time. Because procedures can vary from county to county, you should always seek the advice of the district attorney's office when preparing a search warrant and follow the local rules of returning and filing the warrant.

## REVIEW

1. A search warrant consists of a fill-in-the-blanks portion and a narrative.
2. The Rules of Narrative Writing should be followed in completing the Affidavit or Statement of Probable Cause.
3. Use the first person when introducing yourself to the judge.
4. The description of the premises to be searched must be so good that any police officer can find the location.
5. The property you are looking for must be described with Reasonable Particularity.
6. Use the evidence report format to list the items seized.

## EXERCISES

1. Using the standard of *reasonable particularity,* select and describe a vehicle as if it would be included in a search warrant.
2. Locate the school library and describe it as if it were a premise to be searched.
3. Write an introduction of yourself as if it would be used in a search warrant affidavit.
4. Using the following fact pattern, prepare a search warrant for telephone records. Use the introduction developed in Exercise #3 as a starting point for your affiant introduction. Use today's date and time as reference and assume El Fuego is in the county you live in.

   Two weeks ago, you were hired as a Special Investigator by the California Bureau of Investigations.

   For the past week, you have been assigned to investigate all criminal violations occurring at the Do Drop Inn Bar, 1631 Lager Road, El Fuego.

   Yesterday, while you were undercover in the bar, you saw and heard one of the "regulars" named Billy Biglunch talking sports and horse betting with the owner, Steve Sober. Billy

Biglunch told Steve Sober that he had an inside source at the track and had very good information that a quarter horse named "I'm Not Glue" was on and would win the fifth race easily and that he wanted to bet $100 on the horse to win.

The phone in the bar is on the wall at the east end of the bar. You were sitting on a stool about three feet from it. As soon as Biglunch and Sober finished their conversation, Sober used the phone to call in a bet. You saw him dial 1-213-555-9361 and then say, "This is Sober, I want $100 on 'I'm Not Glue' in the fifth." He hung up the phone and told Biglunch, "Okay, your bet is in through me, that was my bookie and we're all set."

You left the bar and after returning to your office, found that phone service for (213) 555-9361 is provided by the Pacific Telephone Company, 1010 N. Wilshire Blvd., Room #621, Los Angeles, California, and that the number is nonpublished.

## QUIZ

1. How should you describe the location to be searched?
2. What are the three main parts of a search warrant?
   a.
   b.
   c.
3. What two things are generally contained in the search warrant?
   a.
   b.
4. What is the test for describing the property you are looking for?
5. What is another name for the affidavit?
6. What should be included in the beginning of the affidavit?
7. What things should be included in the affiant's introduction?
   a.
   b.
   c.
   d.
   e.
8. How long should an affidavit be?
9. Which part of the warrant contains the list of the evidence seized?
10. What is the suggested order for listing seized evidence?
    a.
    b.
    c.
    d.
    e.

# 9 Dictating Reports

Just as technology has helped solve lots of investigations, it has also improved investigative report writing. Adopting and then applying a successful system of dictating reports from the business world to the criminal justice system has helped streamline the investigative report writing system as well. There are really two parts to the report writing dictation process—dictating and transcribing. Dictating is the part of the system where someone reads or says something that another person transcribes or writes down. Transcribing is the part of the process where a person makes a written copy of the dictated material. While the dictation process has been around the law enforcement community for at least 25 years, it has had mixed results for a number of reasons.

## BENEFITS OF DICTATION

The benefits of a dictation system for law enforcement are many. A prime consideration is that it can save field officers and investigators a lot of time. Almost everyone can talk faster than they can write and when this is multiplied out over many shifts by many officers, the time saved can be substantial. Another benefit is that the finished product is typewritten and usually neater in appearance than a handwritten report. This has a second benefit for those who use the initial reports in that they are easier to read and hopefully make follow-up work easier to do. Problems like being unable to distinguish numbers and letters in license plates, addresses, and phone numbers should be eliminated with an accurately dictated and transcribed report. Another benefit of a solid dictation process is that the system can be connected to an automated record-keeping system and save a lot of time indexing names and addresses. Many departments enter every name and address that appears in a police report into their records bureau master file. This allows investigators to check names at a later date and see every occasion that an officer has been involved with a particular person or address. This is possible by capturing information in various **fields** when the report is being transcribed. The information captured in these fields is automatically entered into the records data base as the report is prepared, thereby saving time and giving field officers up-to-date information on a very timely basis.

Dictation systems are not without problems, however. Officers who have used these systems can give numerous examples of the system not working as advertised. Often cited complaints include the machine running out of tape, equipment shutting off partway through a report, reports including misspelled names or incorrect information, and reports containing very long sentences. A system problem is that there are too many reports waiting to be transcribed, which closes the dictation process to incoming reports. If you are holding several reports to transcribe at the end of your shift, you may have an unpleasant surprise.

## HOW DICTATION SYSTEMS WORK

There are three basic types of dictation systems commonly used in law enforcement. The first is live dictation to a transcriber who collects the information in shorthand and prepares the finished report later. Another version of this type of system is one in which the transcriber types the report into a word processor and the report is completed at that time. A second system requires officers to call into a transcribing center where they are connected to a tape recorder. Transcribers retrieve the tapes, prepare the reports, and transmit them electronically or in hard copy to the originating agency when completed. In the third type of system officers dictate their reports using hand held tape recorders and then deliver the tapes to a transcribing center for completion.

Although problems can occur in any type of transcribing system, there are some things you can do to minimize the likelihood of something bad happening.

1. If you are using a handheld tape recorder, take the time to make sure it works. Batteries can discharge quickly if they are close to their end life. Make sure the recorder is on the correct speed and that the tape is in good condition. Tapes can degrade with multiple uses and compromise the resulting sound quality along with it.

2. Eliminate background noise. If you are dictating from your car, turn the radio down and roll the windows up. Background noise can easily interfere with your voice as you are trying to record, and drown out what you are saying.

3. Allow the system to work for you by giving it enough time to start. If you have ever had your call answered by a voice mail or answering machine that had the beginning of the greeting cut off, you know how difficult it can be to understand what is going on.

4. Give the transcriber good instructions about what you are going to do. Is it a crime report, an arrest report, or a follow-up memo? Be considerate. Also try to estimate the length for the transcriber.

5. Know what you are going to say before you begin. Don't make the report up as you go. Use an outline, or work from your notes, and have the proper format available to refer to. If you are working from one format and the transcriber is working from another, you are asking for problems.

6. Talk slow. It is very difficult to write as fast as most people talk. If you have ever tried to keep up with a college professor or an instructor in an in-service training class, and write their comments word for word, you know how difficult it is. This is especially true if you are using technical terms or phrases that are not common.

7.   Speak clearly and distinctly. Use good pronunciation and pay particular attention to numbers and sound-alike words. Use ZERO when referring to the number and OH when referring to the letter O. Interchanging these two can create a lot of difficulty when trying to track license plates that have been improperly recorded. You should also spell all names if there is any chance for confusion. Never give a name and say common spelling if there is a chance the transcriber will interpret it differently than you intend. Words that sound alike are also potential problems, for example, THRU and THROUGH, WAIT and WEIGHT, PASSED and PAST. When in doubt, spell it out.

8.   Specify the grammar and punctuation that is needed. For example, *I saw Brown, capital B-R-O-W-N, hit Smith, capital S-M-I-T-H period, new sentence, I went to the hospital and spoke to Brown, capital B-R-O-W-N, etc.* If you are quoting someone say: *Brown, capital B-R-O-W-N, said comma quotation mark I did it period end quotation mark.* Learning to call out and specify the grammar and punctuation is a real challenge for many officers.

9.   Stay calm and recognize that mechanical failures and other things are going to happen from time to time. It is part of the world you live in and not the transcribers fault. Getting frustrated will not solve a thing.

10.  Because mechanical failures are going to happen, save your notes until you are sure the report has been completed. If you have to dictate a report a second time because a tape was misplaced or broke, having your notes will be a real help. You may not see the finished report for days or weeks if it is approved and routed through the system, so keeping notes available is a wise thing to do.

11.  Stay focused on what you are doing and do not talk to others while you are dictating. This can be very confusing for a transcriber who is trying to figure out if you are talking as part of the report or to someone who is not involved.

Although dictation can be a positive tool for some officers, it is not for everyone. You should have a solid writing background before you begin dictating reports. When it comes to report writing there is no substitute for good training and experience. This combination is an excellent foundation to build advanced writing skills on. Most police officers begin their report writing training with academy instruction followed by field experience hand writing reports under the supervision of a training officer. Very few officers receive training in this area during the academy and the skills they acquire are usually learned on the job by watching others and modeling their behaviors. This is an area of your training that you may have to take responsibility for and be in charge of learning the proper way to dictate. Be open to constructive criticism when you begin. Talking to transcribers and learning firsthand about the problems they experience is a good step to take when you are learning how to dictate reports. These same transcribers will probably be able to suggest ways to correct the problems. Ask for a tape of a well-dictated report and listen to it a few times. Pay particular attention to the speed at which the officer is talking. Listen for the grammar directions and the pronunciation used, and then practice before you start dictating. If you find that this report writing tool works for you, your daily report writing time should decrease.

Last but not least, remember that a successful dictation system is a team effort. You are part of the team and the transcriber typing your reports is an equal part of the team. Remember to use common courtesy when setting up the tapes and giving directions about what you would like to have done. Dictating reports is not an easy task for many officers. It is a learned skill with a steep learning curve. The two expressions that are most under used when dictating reports are PLEASE and THANK YOU. They are also the two expressions that can make a transcribing team achieve greatness.

## REVIEW

1. It takes two or more people to have an effective dictation process.
2. Good equipment is key to a successful system.
3. Background noise can be a major problem for transcribers.
4. Have a plan and an outline before you start dictating.
5. Speak slowly and clearly.
6. Systems sometimes fail, so keep your notes.
7. Specify punctuation and grammar.
8. Use common courtesy and respect the transcribers.

## EXERCISES

1. Develop a one-page theft report and complete a handwritten report using a crime report face sheet. With this report as a guide, dictate the report using a handheld tape recorder. Listen to the tape and note your pronunciation, how fast you talked, and the directions you gave. Is your work understandable?
2. Develop a cheat sheet to use for dictating reports. Now repeat the first exercise and give the tape to a classmate to evaluate. Do you still have room for improvement?
3. Dictate a report and give it to a classmate to transcribe. The finished product should show exactly what was on the tape. Any differences between what you said and what you intended?

## QUIZ

1. Dictation has been a plus for law enforcement report writing.
   a. True
   b. False
2. Give two examples of how dictation can benefit the report writing process.
   a.
   b.
3. What are some of the problems with dictation systems?
   a.
   b.
   c.
4. Give examples of two types of dictation systems.
   a.
   b.
5. What is a common reason handheld tape recorders fail?
6. Why is background noise an issue with dictated reports?
7. You should avoid giving the transcribers directions about your report because it takes up too much tape.
   a. True
   b. False
8. Why is it a good idea to spell out unusual or like sounding words?
9. Transcribers are able to distinguish between what you are dictating as part of the report and what you are saying to someone nearby.
   a. True
   b. False
10. Moving from handwritten reports to transcribed reports is an easy task for most officers.
    a. True
    b. False

# Index